Probes and Tags to Study Biomolecular Function

Edited by
Lawrence W. Miller

Related Titles

Schreiber, S. L., Kapoor, T., Wess, G. (eds.)

Chemical Biology

From Small Moleculesto Systems Biology and Drug Design

2007

ISBN: 978-3-527-31150-7

Daunert, S., Deo, S. K. (eds.)

Photoproteins in Bioanalysis

2006

ISBN: 978-3-527-31016-6

Fielding, C. J. (ed.)

Lipid Rafts and Caveolae

From Membrane Biophysics to Cell Biology

2006

ISBN: 978-3-527-31261-0

Chalfie, M., Kain, S. (eds.)

Green Fluorescent Protein

Properties, Applications and Protocols

2005

ISBN: 978-0-471-73682-0

Tamm, L. K. (ed.)

Protein-Lipid Interactions

From Membrane Domains to Cellular Networks

2005

ISBN: 978-3-527-31151-4

Probes and Tags to Study Biomolecular Function

for Proteins, RNA, and Membranes

Edited by
Lawrence W. Miller

WILEY-VCH Verlag GmbH & Co. KGaA

The Editor

Prof. Dr. Lawrence W. Miller
Department of Chemistry
University of Illinois
845 West Taylor Street
Chicago, IL 60607-7061
USA

All books published by Wiley-VCH are carefully produced. Nevertheless, authors, editors, and publisher do not warrant the information contained in these books, including this book, to be free of errors. Readers are advised to keep in mind that statements, data, illustrations, procedural details or other items may inadvertently be inaccurate.

Library of Congress Card No.: applied for

British Library Cataloguing-in-Publication Data
A catalogue record for this book is available from the British Library.

Bibliographic information published by the Deutsche Nationalbibliothek
Die Deutsche Nationalbibliothek lists this publication in the Deutsche Nationalbibliografie; detailed bibliographic data are available in the Internet at http://dnb.d-nb.de.

© 2008 WILEY-VCH Verlag GmbH & Co. KGaA, Weinheim

All rights reserved (including those of translation into other languages). No part of this book may be reproduced in any form – by photoprinting, microfilm, or any other means – nor transmitted or translated into a machine language without written permission from the publishers. Registered names, trademarks, etc. used in this book, even when not specifically marked as such, are not to be considered unprotected by law.

Typesetting Thomson Digital, Noida, India
Printing betz-druck GmbH, Darmstadt
Binding Litges & Dopf Buchbinderei GmbH, Heppenheim
Cover Design Grafik-Design Schulz, Fußgönheim

Printed in the Federal Republic of Germany
Printed on acid-free paper

ISBN: 978-3-527-31566-6

Contents

Preface *XI*
List of Contributors *XIII*

1 Fluorescent Sterols for the Study of Cholesterol Trafficking in Living Cells *1*
Avery L. McIntosh, Huan Huang, Barbara P. Atshaves, Stephan M. Storey, Adalberto M. Gallegos, Thomas A. Spencer, Robert Bittman, Yoshiko Ohno-Iwashita, Ann B. Kier, and Friedhelm Schroeder
1.1 Introduction *1*
1.2 Methods for Imaging Fluorescent Sterols in Living Cells: Confocal and Multiphoton Laser Scanning Microscopy *3*
1.2.1 Sources of Cholesterol and Fluorescent Sterols *3*
1.2.2 Spectral Properties of Fluorescent Sterols *3*
1.2.3 Fluorescent Sterol Labeling Methodology *3*
1.2.3.1 Method 1: Direct Labeling *3*
1.2.3.2 Method 2: Fluorescent Sterol Incorporation by Large Unilamellar Vesicles *5*
1.2.3.3 Method 3: Fluorescent Sterol-methyl-β-cyclodextrin (FS-MβCD) Complexes *5*
1.2.3.4 Method 4: Fluorescent Labeling of High Density Lipoproteins (HDL) *5*
1.2.3.5 Method 5: BCθ *6*
1.2.4 Confocal Laser Scanning Microscopy (CLSM) and Multiphoton Laser Scanning Microscopy (MPLSM) of Sterol Probes *6*
1.3 Cholesterol Structure and Distribution in Membranes *7*
1.4 NBD-Cholesterol *8*
1.4.1 22-NBD-Cholesterol *9*
1.4.2 25-NBD-Cholesterol *11*
1.5 Dansyl-Cholesterol *13*
1.6 BODIPY-Cholesterol *14*

Probes and Tags to Study Biomolecular Function. Lawrence W. Miller (Ed.)
Copyright © 2008 WILEY-VCH Verlag GmbH & Co. KGaA, Weinheim
ISBN: 978-3-527-31566-6

1.7	Dehydroergosterol (DHE) 17
1.8	22-(p-Benzoylphenoxy)-23,24-bisnorcholan-5-en-3β-ol (FCBP) Photoactivatable Sterol 20
1.9	BCθ 20
1.10	Conclusion 23
	References 25

2	**Lipid Binding Proteins to Study Localization of Phosphoinositides** 35
	Guillaume Halet and Patricia Viard
2.1	Introduction: Phosphoinositide Signaling 35
2.2	Monitoring PI Distribution and Dynamics 37
2.2.1	Detection of PI Species Using Antibodies 37
2.2.2	Fluorescent PI Derivatives 38
2.2.3	Fluorescent PI-Binding Domains 39
2.2.3.1	Choosing the Right Domain 39
2.2.3.2	Imaging PI Probes 40
2.3	Detection of PtdIns(3,4,5)P_3 Synthesis in Transfected Mammalian Cells 41
2.3.1	Transfection with Plasmid DNA Encoding GFP-PH$_{GRP1}$ 41
2.3.2	Detection of PtdIns(3,4,5)P_3 Synthesis by a Constitutively-active PI3K 42
2.3.3	Detection of PtdIns(3,4,5)P_3 Synthesis after Stimulation with EGF 43
2.4	Monitoring PtdIns(4,5)P_2 Dynamics in Mouse Oocytes 43
2.4.1	Making cRNAs 44
2.4.2	PtdIns(4,5)P_2 Dynamics in Mouse Oocytes at Fertilization and after Treatment with Ionomycin 45
2.5	Limitations of the Technique 45
2.5.1	PI Probes may Detect Only a Subset of the Total PI Pool 46
2.5.2	PI Probes can Interfere with Normal Cellular Function 48
2.5.3	Binding of PI Probes to Inositol Phosphates 48
2.6	Conclusion 48
	References 49

3	**The Use of Lipid-Binding Toxins to Study the Distribution and Dynamics of Sphingolipids and Cholesterol** 53
	Reiko Ishitsuka and Toshihide Kobayashi
3.1	Introduction 53
3.2	Cholera Toxin 54
3.2.1	Introduction 54
3.2.2	CTB as a Tool to Study Cell Surface Lipid Rafts 55
3.2.2.1	Use of CTB for Biophysical Characterization of Lipid Rafts 55
3.2.2.2	Electron Microscopic Studies Using CTB 56
3.2.3	Intracellular Trafficking of CT 56
3.2.4	Protocols 57

3.2.4.1	Materials	57
3.2.4.2	Observation of Trafficking of GM1 in Living Cells Using CTB	57
3.3	Lysenin	59
3.3.1	Introduction	59
3.3.2	Lysenin Binds Clustered Sphingomyelin	59
3.3.3	Non-Toxic Lysenin as a Sphingomyelin Probe	60
3.3.4	Protocols	62
3.3.4.1	Materials	62
3.3.4.2	Cellular Staining of Sphingomyelin by Lysenin	62
3.3.4.3	Observation of Sphingomyelin on the Plasma Membrane of Living Cells Using Non-Toxic Lysenin	63
3.4	Perfringolysin O	63
3.4.1	Introduction	63
3.4.2	Non-Toxic Derivatives of PFO Bind Cholesterol-Rich Domains	64
3.4.3	Use of BCθ to Detect Cholesterol-Rich Domains	64
3.4.4	Protocol	65
3.4.4.1	Materials	65
3.4.4.2	Staining of Cholesterol-Rich Domain in the Plasma Membrane Using BCθ	65
3.5	Aerolysin	66
3.5.1	Introduction	66
3.5.2	Use of Aerolysin to Detect GPI-Anchored Proteins	66
	References	67
4	**"FlAsH" Protein Labeling**	**73**
	Stefan Jakobs, Martin Andresen, and Christian A. Wurm	
4.1	Introduction	73
4.1.1	The Biarsenical-Tetracysteine System	74
4.1.2	Improved TetCys Motifs	75
4.1.3	Fluorescent Biarsenical Ligands	75
4.1.4	Applications of the Biarsenical-Tetracysteine System	76
4.1.5	Staining in Various Model Organisms	78
4.1.6	FlAsH Labeling in *S. cerevisiae*	79
4.1.7	Outlook	81
4.2	Use of the Biarsenical-Tetracysteine System in *S. cerevisiae*	82
4.2.1	Materials	82
4.2.1.1	Growth Media	82
4.2.1.2	Buffers (Required Stock Solutions)	82
4.2.2	Labeling Protocols	83
4.2.2.1	Overnight Staining	83
4.2.2.2	Pulse Staining	84
4.2.2.3	Mounting and Microscopy	84
4.3	Short Protocols	85
4.3.1	Overnight Staining	85
4.3.2	Pulse Staining	85

4.4	Troubleshooting 86	
	References 88	
5	**AGT/SNAP-Tag: A Versatile Tag for Covalent Protein Labeling** 89	
	Arnaud Gautier, Kai Johnsson, and Helen O'Hare	
5.1	Introduction 89	
5.2	Labeling SNAP-Tag Fusion Proteins with BG Derivatives 90	
5.2.1	Human O^6-Alkylguanine-DNA Alkyltransferase 90	
5.2.2	The Principle of SNAP-Tag Labeling 91	
5.3	SNAP-Tag for Cell Imaging 95	
5.4	Procedures for SNAP-Tag Labeling 97	
5.4.1	Standard Protocol for Fluorescent Imaging of SNAP-Tagged Proteins in Transiently Transfected Adherent Mammalian Cell Culture 97	
5.4.2	Technical Notes 99	
5.4.2.1	Counterstaining or Fixing 99	
5.4.2.2	Photobleaching 99	
5.4.2.3	Checking Expression of the Fusion Protein 99	
5.4.2.4	Labeling AGT *in vitro* 100	
5.4.2.5	Pulse-Chase Labeling 100	
5.4.2.6	Labeling on the Cell Surface Using Non-Permeable Dyes (BG-FL, BG-Cy3, BG-Cy5) 100	
5.4.2.7	Labeling Intracellular Proteins Using Non-Permeable Dyes 101	
5.4.2.8	Multicolor Labeling of More Than One Protein 101	
5.4.2.9	Measuring Protein–Protein Interactions or Conformational Changes by FRET 101	
5.4.2.10	Sensors 102	
5.5	Broader Applications of SNAP-Tag to Study Protein Function 102	
5.5.1	Induction of Protein Dimerization by Covalent Labeling in Living Cells 102	
5.5.2	SNAP-tag-Mediated Covalent Immobilization of Fusion Proteins on BG-Functionalized Surfaces 104	
5.6	Conclusion 105	
	References 106	
6	**Trimethoprim Derivatives for Labeling Dihydrofolate Reductase Fusion Proteins in Living Mammalian Cells** 109	
	Lawrence W. Miller and Virginia W. Cornish	
6.1	Introduction 109	
6.2	Preparation of *E. coli* Expression Vectors 112	
6.2.1	Materials 112	
6.2.2	Plasmids for Over-Expression of *E. coli* DHFR Fusion Proteins in Mammalian Cells 112	
6.2.2.1	Nucleus-Localized eDHFR Expression Plasmid (pLM1302). Construction of Nucleus-Localized eDHFR Vector (Plasmid pLM1302) 112	

6.2.2.2	Myosin Light Chain Kinase eDHFR Plasmid *112*
6.3	Synthesis of Fluorescent Trimethoprim Derivatives *113*
6.4	Cell Growth and Transfection *114*
6.5	Protein Labeling and Microscopy *114*
6.6	Results and Discussion *115*
6.7	Conclusion *117*
	References *118*

7	**Phosphopantetheinyl Transferase Catalyzed Protein Labeling and Molecular Imaging** *121*
	Norman J. Marshall and Jun Yin
7.1	Introduction *121*
7.2	Protein Posttranslational Modification by Phosphopantetheinyl Transferases *123*
7.3	Protein Labeling Via Carrier Protein Fusions *127*
7.4	Orthogonal Protein Labeling by Short Peptide Tags *131*
7.5	Summary and Perspectives *133*
	References *134*

8	**Bioorthogonal Chemical Transformations in Proteins by an Expanded Genetic Code** *139*
	Birgit Wiltschi and Nediljko Budisa
8.1	Introduction *139*
8.2	Chemical Transformations at the Protein N-terminus Classical Approaches *140*
8.2.1	Biomimetic Transamination *140*
8.2.2	Enzymatic Modification of the N-terminus *143*
8.3	Chemical Transformations Using an Expanded Genetic Code *145*
8.3.1	The Genetic Code and Its Expansion *145*
8.3.2	Cell-free N-terminal Labeling with Modified Initiator tRNA *147*
8.3.3	Cell-free N-terminal Labeling with Suppressor Initiator tRNA *149*
8.4	Modifications Internal to Proteins *149*
8.5	Chemical Transformations at the Protein C-terminus *155*
8.6	Summary and Outlook *158*
	References *159*

9	**Using the Bacteriophage MS2 Coat Protein–RNA Binding Interaction to Visualize RNA in Living Cells** *163*
	Jeffrey A. Chao, Kevin Czaplinski, and Robert H. Singer
9.1	Introduction *163*
9.2	Construction of an MBS-Containing Reporter RNA *165*
9.3	Construction of an MS2 CP-FP Chimera *166*
9.4	Co-introduction of MS2 Reporter RNA and MS2 CP-FP *168*
9.5	Microscopy Platform for Single Molecule Detection of RNAs in Living Cells *168*

9.5.1	Components of the Imaging System	*169*
9.6	Protocols for Co-expressing MS2 CP-FP- and MBS-Containing Plasmids for Live Cell Imaging	*170*
9.7	Image Analysis and Quantification of mRNA Molecules	*171*
	References *172*	

Index *175*

Preface

A central challenge of biochemistry and molecular cell biology is to understand how biomolecules interact and react to organize and control cell growth, structure and function. The contributions in this book reveal many ways in which the technologies of specific chemical and genetic labeling can be used in conjunction with optical microscopy to dynamically analyze the spatial and temporal organization of proteins, lipids, and even messenger RNA in single living cells.

More than half a century of progress in biochemistry, genetics, molecular biology and cellular physiology has yielded a rich understanding of the macromolecular basis of structure and information flow in living systems. Complete genomes have been sequenced, tens of thousands of protein structures have been determined, and the activities and functions of thousands of enzymes have been analyzed. Over the past decade, methods that allow for the real-time analysis of molecular function in the environment of the living cell or organism have come to prominence. The ability to dynamically and non-destructively image the translocations, interactions or reactions of one or more chemically unique biomolecules affords a mechanistic understanding of cellular function that is not accessible using traditional biochemical assays.

Selectivity is the unifying idea that runs throughout this book's chapters. *In vivo* studies require non-invasive methods of imparting unique optical or chemical functionalities to particular molecular species. One strategy is to prepare a soluble probe molecule that is fluorescent or otherwise detectable, but which retains the same cellular localization and biological function of the unlabeled species. The other general strategy is to genetically label a protein with a functional tag. This tag can be one of many commonly used autofluorescent proteins, or it can be a receptor protein or polypeptide that binds to a soluble, cell-permeable, small molecule that has the desired functionality.

The first three contributions in this volume describe various strategies for probing the function of the lipid bilayer. The spatiotemporal dynamics of phosphoinositides and their role in activating signaling cascades or membrane trafficking events can be microscopically visualized by expressing fluorescent proteins fused to phosphoinositide-binding motifs. The organization of cholesterol and sphingolipids

Probes and Tags to Study Biomolecular Function. Lawrence W. Miller (Ed.)
Copyright © 2008 WILEY-VCH Verlag GmbH & Co. KGaA, Weinheim
ISBN: 978-3-527-31566-6

into discrete microdomains within the lipid bilayer can be studied with two approaches described herein. Naturally occurring and synthetic fluorescent analogues of cholesterol can be incorporated directly into the lipid bilayer of living cells, allowing microscopic visualization of cholesterol-rich microdomains. Alternatively, lipid-binding toxins can be used to label and detect sphingolipids, cholesterol and GPI-anchored proteins.

Several contributors offer experimental strategies for appending specific proteins with synthetically optimized small molecules. Proteins of biological interest can be genetically encoded as fusions to receptor proteins or polypeptides. Cell-permeable ligands can be synthesized with a variety of functionalities, including enhanced fluorescence, photoaffinity or analyte sensing capability. Upon addition to culture medium, these ligands can bind specifically and stably to the fusion protein chimeras via high-affinity non-covalent interaction or enzyme-mediated covalent linkage. The smallest possible functional tag for a protein is a single amino acid. A comprehensive description of methods for incorporating unnatural amino acids with specific chemical or optical properties into proteins, both *in vitro* and *in vivo*, is included.

There are comparatively few robust methods for the sequence-specific labeling of nucleic acid species in living cells or organisms. Fusion of bacteriophage MS2 coat protein to fluorescent proteins allows labeling of messenger RNA containing the cognate MS2 hairpin binding site. Tagged mRNAs can be microscopically visualized in living mammalian cells with single transcript resolution. A thorough description of this technique is provided in the final chapter of this volume.

The contributions to *Probes and Tags to Study Biomolecular Function* reflect the efforts of chemists and biologists to bring the study of biochemistry from the test tube to the living cell. Taken together, the experimental tools described in this volume reflect the state-of-the-art in technologies to label biomolecules for *in vivo* studies. The step-by-step protocols and illustrations of typical applications will enable researchers to select the best solution for their experimental problems. Looking ahead, the real-time analysis of molecular function within living cells will continue to be an indispensable approach to studying mechanistic biology. Our contributors have laid a foundation for the future development of more robust and selective labeling technologies that should allow for *in vivo* detection with even greater sensitivity and spatio-temporal resolution.

Chicago, 2008

Lawrence W. Miller

List of Contributors

Martin Andresen
Max Planck Institute for Biophysical
Chemistry
Mitochondrial Structure and Dynamics
Group
Department of NanoBiophotonics
Am Fassberg 11
37077 Goettingen
Germany

Barbara P. Atshaves
Texas A&M University
Department of Physiology and
Pharmacology
TVMC
College Station
TX 77843-4466
USA

Robert Bittman
Queens College
Department of Chemistry and
Biochemistry
CUNY
Flushing
NY 11367-1597
USA

Nediljko Budisa
Max Planck Institute of Biochemistry
BioFuture Independent Research Group
Molecular Biotechnology
Am Klopferspitz 18
82152 Martinsried
Germany

Jeffrey A. Chao
Albert Einstein College of Medicine
Department of Anatomy and Structural
Biology
1300 Morris Park Avenue
Bronx NY 10461
USA

Virginia W. Cornish
Columbia University
Department of Chemistry
3000 Broadway
MC 3111
New York
NY 10027
USA

Kevin Czaplinski
Albert Einstein College of Medicine
Department of Anatomy and Structural
Biology
1300 Morris Park Avenue
Bronx NY 10461
USA

List of Contributors

Adalberto M. Gallegos
Texas A&M University
Department of Pathobiology
TVMC, College Station
TX 77843-4466
USA

Arnaud Gautier
École Polytechnique Fédérale de Lausanne
Institute of Chemical Sciences and Engineering
1015 Lausanne
Switzerland

Guillaume Halet
University College London
Department of Physiology
Gower Street
London WC1E 6BT
UK

Huan Huang
Texas A&M University
Department of Physiology and Pharmacology
TVMC
College Station
TX 77843-4466
USA

Reiko Ishitsuka
RIKEN Discovery Research Institute
Lipid Biology Laboratory
2-1, Hirosawa
Wako-shi
Saitama 351-0198
Japan

Stefan Jakobs
Max Planck Institute for Biophysical Chemistry
Mitochondrial Structure and Dynamics Group
Department of NanoBiophotonics
Am Fassberg 11
37077 Goettingen
Germany

Kai Johnsson
École Polytechnique Fédérale de Lausanne
Institute of Chemical Sciences and Engineering
1015 Lausanne
Switzerland

Ann B. Kier
Texas A&M University
Department of Pathobiology
TVMC, College Station
TX 77843-4466
USA

Toshihide Kobayashi
RIKEN Discovery Research Institute
Lipid Biology Laboratory
2-1, Hirosawa
Wako-shi
Saitama 351-0198
Japan

Norman J. Marshall
The University of Chicago
Department of Chemistry
929 E. 57th Street
GCIS 505A
Chicago
IL 60637
USA

List of Contributors

Avery L. McIntosh
Texas A&M University
Department of Physiology and
Pharmacology
TVMC
College Station
TX 77843-4466
USA

Lawrence W. Miller
University of Illinois at Chicago
Department of Chemistry
845 W. Taylor Street
MC 111
Chicago
IL 60607
USA

Helen O'Hare
École Polytechnique Fédérale de
Lausanne
Institute of Chemical Sciences and
Engineering
1015 Lausanne
Switzerland

Yoshiko Ohno-Iwashita
Tokyo Metropolitan Institute of
Gerontology
Biomembrane Research Group
Tokyo 173-0015
Japan

Friedhelm Schroeder
Texas A&M University
Department of Physiology and
Pharmacology
TVMC
College Station
TX 77843-4466
USA

Robert H. Singer
Albert Einstein College of Medicine
Department of Anatomy and Structural
Biology
1300 Morris Park Avenue
Bronx NY 10461
USA

Thomas A. Spencer
Dartmouth College
Department of Chemistry
Hanover
NH 03755-1812
USA

Stephan M. Storey
Texas A&M University
Department of Physiology and
Pharmacology
TVMC
College Station
TX 77843-4466
USA

Patricia Viard
University College London
Department of Pharmacology
Gower Street
London WC1E 6BT
UK

Christian A. Wurm
Max Planck Institute for Biophysical
Chemistry
Mitochondrial Structure and Dynamics
Group
Department of NanoBiophotonics
Am Fassberg 11
37077 Goettingen
Germany

Birgit Wiltschi
Max Planck Institute of Biochemistry
BioFuture Independent Research Group
Molecular Biotechnology
Am Klopferspitz 18
82152 Martinsried
Germany

Jun Yin
The University of Chicago
Department of Chemistry
929 E. 57th Street
GCIS 505A
Chicago
IL 60637
USA

1
Fluorescent Sterols for the Study of Cholesterol Trafficking in Living Cells

Avery L. McIntosh, Huan Huang, Barbara P. Atshaves, Stephan M. Storey, Adalberto M. Gallegos, Thomas A. Spencer, Robert Bittman, Yoshiko Ohno-Iwashita, Ann B. Kier, and Friedhelm Schroeder

1.1
Introduction

Cholesterol is one of the most important constituents in lipid microdomain (also known as lipid raft) formation in model [1] and plasma membranes [2]. Plasma membrane lipid rafts are not only enriched in cholesterol but also sphingolipids and phospholipids with saturated and monounsaturated fatty acyl chains [3]. The physical state of these microdomains is a distinct, liquid-ordered PM phase, intermediate between the fluid liquid-crystalline and rigid gel phases [3–15]. The existence, properties, and regulation of lipid rafts (reviewed in [3, 16–18]) provide a framework for the study of the location and function of membrane protein receptors, transporters, and downstream signaling molecules (Figure 1.1) that regulate uptake of cholesterol [19–42], fatty acids [43–45], glucose [19, 46–57], and other related processes [3, 21, 58–65].

Membrane lipid microdomains are a source of continuing complexity and frustration, as evidenced by the titles bestowed on recent reviews [16, 17, 65–75]. The range of sizes reported for lipid microdomains vary from the very small (1–10 nm) to the very large (>1 μm) [74, 76–79]. The small can be described as single annular shells of cholesterol surrounding lipids or proteins, that are refractory to biochemical isolation [74, 76, 80] while the larger domains, such as whole microvilli, filopodia, pseudopodia, basolateral, apical areas and bile canaliculi, are subject to biochemical isolation from tissues rich in polarized cells (liver, intestine, kidney) [79, 81–86].

The smallest microdomains can be localized within larger domains [87, 88], as elegantly shown for microvilli [89] but, in large part, intermediate-sized microdomains, 50–700 nm, have been the subject of most studies. Although morphologists observed the first intermediate-sized microdomains (50–100 nm), now termed caveolae, over 50 years ago [90, 91], their functional significance was uncertain until

Probes and Tags to Study Biomolecular Function. Lawrence W. Miller (Ed.)
Copyright © 2008 WILEY-VCH Verlag GmbH & Co. KGaA, Weinheim
ISBN: 978-3-527-31566-6

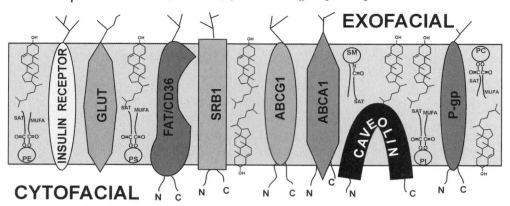

Figure 1.1 Cholesterol-rich microdomains within cellular plasma membrane. Membrane bilayer shows receptors and transport/translocase proteins distributed in lipid microdomains enriched in cholesterol.

the discovery of caveolin-1, the protein directing formation of this remarkably stable subclass of microdomain [17, 20, 36, 59, 61, 74, 75, 92, 93]. While caveolin provided the first convenient marker for biochemical isolation of caveolae, identical fractionation techniques also isolated similar size microdomains from cells lacking caveolin [20], indicative of "caveolae" as a subset of "lipid rafts" [94, 95]. Recently, emergence of many other types of "lipid rafts" has led to the appreciation that multiple intermediate-sized microdomains may coexist within the plasma membrane and likely in subcellular membranes as well. To clarify the growing complexity, a provisional definition of "lipid rafts" emerged at the 2006 Keystone Symposium: "Membrane rafts are small (10–200 nm), heterogeneous, highly dynamic, cholesterol- and sphingolipid-enriched domains that compartmentalize cellular processes. Small rafts can sometimes be stabilized to form larger platforms through protein–protein and protein–lipid interactions [18]." Lipid raft quantity (none to nearly the entire membrane [96]) and purity vary depending on the isolation techniques, with those obtained as detergent-resistant membranes (DRM) or high pH carbonate buffer-isolated membranes containing as much as 30–75% cross-contaminating non-raft and intracellular proteins [3, 4, 13, 16, 17, 68, 96–115]. DRMs can contain artifactual sterol structure (crystalline) and exhibit abnormal sterol efflux unresponsive to intracellular cholesterol transfer proteins [116].

Despite the obvious importance of cholesterol for lipid raft formation, very little is known regarding the actual properties of cholesterol itself within these microdomains. The lack of understanding of cholesterol's properties in microdomains underlines the importance of discovering and utilizing fluorescent sterol probes in living cells to study cholesterol trafficking, plasma membrane cholesterol organization (lateral, transbilayer, polar interface, order), dynamic interactions of cholesterol with lipids and proteins, and cholesterol transport in caveolae/lipid rafts [1, 10, 14, 17, 65, 117–119].

1.2
Methods for Imaging Fluorescent Sterols in Living Cells: Confocal and Multiphoton Laser Scanning Microscopy

1.2.1
Sources of Cholesterol and Fluorescent Sterols

Cholesterol is available from several commercial sources, e.g., AvantiPolar Lipids, (Alabaster, AL) (Figure 1.2A); 22-NBD-cholesterol or 22-(N-(7-nitrobenz-2-oxa-1, 3-diazol-4-yl)amino)-23,24-bisnor-5-cholen-3β-ol (Figure 1.2C) from Invitrogen (Carlsbad, CA), 25-NBD-cholesterol or 25-(N-[(7-nitrobenz-2-oxa-1,3-diazol-4-yl)-methyl]amino)-27-norcholesterol (Figure 1.2E) from AvantiPolar Lipids (Alabaster, AL). DChol or dansyl-cholesterol (Figure 1.2D) was synthesized as described [120]. BODIPY-Chol 2 with (4,4-difluoro-4-bora-3a,4a-diaza-s-indacene) located in the aliphatic tail of cholesterol (Figure 1.2F) was synthesized as described earlier [121]. DHE or dehydroergosterol (Figure 1.2B) is available from Sigma (St. Louis, MO) or was synthesized as described earlier [122–124]. FCBP or 22-(p-benzoylphenoxy)-23,24-bisnorcholan-5-en-3b-ol (Figure 1.2G) was synthesized as described earlier [125]. BCθ was derived by biotinylation of the protease-nicked θ-toxin [126, 127]. Purity of some fluorescent sterols was determined through the use of absorption spectroscopy, HPLC, and APCI-mass spectroscopy [128, 129]. Other sterols were accepted as per commercial sources or as cited in published papers.

1.2.2
Spectral Properties of Fluorescent Sterols

A Cary 100 (Varian, Palo Alto, CA) was used to acquire fluorophore absorbance spectra. Typical scans were acquired over the wavelength range, 200–600 nm, using quartz cuvettes. The fluorescence excitation and emission spectra of the fluorophores were obtained in quartz cuvettes with a Cary Eclipse spectrofluorometer (Varian, Palo Alto, CA) using a pulsed xenon lamp and analog detection mode. Emission and excitation monochromator slits were set at 5 nm spectral bandwidth.

1.2.3
Fluorescent Sterol Labeling Methodology

1.2.3.1 Method 1: Direct Labeling
Stock solutions were made by dissolving the fluorescent sterol in ethanol which was subsequently stored under N_2 at −80 °C. Intact L-cell fibroblasts were cultured as described previously in serum-containing medium on LabTek two-chambered coverglasses (VWR, Sugarland, TX) overnight [128]. After overnight growth, the medium was removed. For DHE and dansyl cholesterol the cells were incubated with the probe in serum-containing medium at 37 °C for times dependent upon the particular sterol and maintained at 5% CO_2 and subsequently imaged in PBS using laser scanning

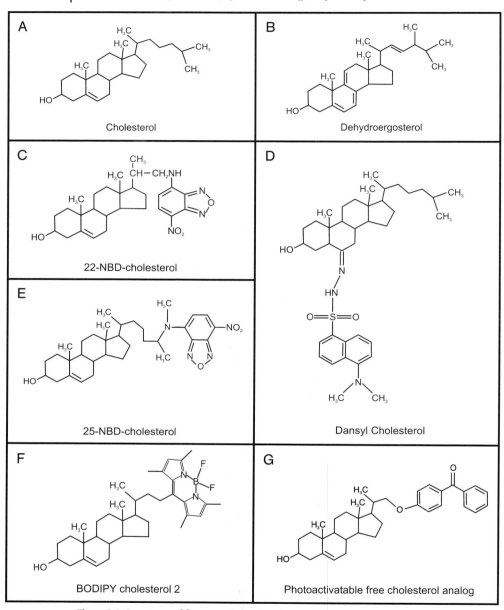

Figure 1.2 Structures of fluorescent sterol analogs.
(A) cholesterol (B) dehydroergosterol (C) 22-NBD-cholesterol
(D) Dansyl-cholesterol (6-dansyl-cholestanol) (E) 25-NBD-cholesterol (F) BODIPY-cholesterol-2 (G) FCBP photoactivatable sterol cross-linker.

microscopy as described below. In the case of the two moieties of NBD-cholesterol, the cells were washed with PBS and the fluorescent sterol was incubated in PBS while the resultant uptake was monitored by confocal laser scanning microscopy. For all sterols, the fluorescent sterol was added from an ethanolic stock solution so that the amount of ethanol was less than 0.2%.

1.2.3.2 Method 2: Fluorescent Sterol Incorporation by Large Unilamellar Vesicles

As described previously, large unilamellar vesicles (LUVs) were prepared by mixing 1-palmitoyl-2-oleoyl-sn-glycero-3-phosphocholine (POPC), 1,2-dioleoyl-sn-glycero-3-[phospho-L-serine] (DOPS) (Avanti Polar Lipids, Alabaster, AL), and the fluorescent probe from stock solutions in chloroform in the proportions 65 mol% POPC, 35 mol% sterol, and 10 mol% DOPS unless otherwise stated [128]. The lipid mixture was first dried under N_2 and then *in vacuo* overnight. Then 10 mM PIPES (1,4-piperazinediethanesulfonic acid) buffer at pH 7.2 was added, followed by sonication for 10 min. The multi-lamellar vesicles were then extruded using a Mini-Extruder (Avanti Polar Lipids, Alabaster, AL), producing large unilamellar vesicles with median radii 53 ± 10 nm. L-cells were cultured on chambered coverglasses in serum-containing media and then washed three times with PBS. Subsequently, an aliquot of LUVs was added to the L-cells in PBS so that the cells were incubated overnight with 20 µg of sterol, except in the case of DHE where the cells were incubated with LUVs in serum-containing media. Prior to imaging the cells were washed three times with PBS and then imaged in PBS.

1.2.3.3 Method 3: Fluorescent Sterol-methyl-β-cyclodextrin (FS-MβCD) Complexes

The fluorescent sterol-MβCD complex was prepared in a 1 : 3 ratio of fluorescent sterol : MβCD by mixing the fluorescent sterol as a powder in an aqueous solution of MβCD. The vessel was filled with N_2 and continuously vortexed under light protection for 24 h at room temperature. The solution was filtered through a 0.2 µm filter to remove insoluble material and large aggregates before use. The fluorescent sterol-MβCD complex was then added to the cells and incubated for 45 min at room temperature in PBS. The cells were subsequently washed three times with PBS and imaged by laser scanning microscopy in PBS.

1.2.3.4 Method 4: Fluorescent Labeling of High Density Lipoproteins (HDL)

A previously published protocol [130] was modified for the fluorescent labeling of high density lipoproteins. The lipids, 2.3 µmol egg phosphatidylcholine, 0.6 µmol sphingomyelin, 2.24 µmol fluorescent sterol, 0.6 µmol cholesterol, and 0.34 µmol triolein, were mixed in chloroform, dried under nitrogen, and then *in vacuo* overnight. 5 ml of a buffer composed of 10 mM Tris-HCl, pH 8.0, 150 mM NaCl, and 0.25 mM EDTA was added and the mixture was vortexed and sonicated at 52 °C for 40 min. 5 ml of HDL in 1 ml of a new buffer (addition of 2.5 M urea in the previous buffer), was added drop-wise over 5 min at 42 °C and sonicated for 10 min at 52 °C. The supernatant was collected after centrifugation and centrifuged again using a sucrose gradient. Neighboring fractions containing the highest amounts of lipoproteins and the fluorescent sterol (determined by

Western blotting and fluorescence emission spectra) were consolidated for cell labeling.

Primary murine hepatocytes were cultured on collagen-coated Nalge Nunc Lab-Tek two-chamber coverglasses overnight in complete media at 37 °C with 5% CO_2. The hepatocytes were washed twice with warm PBS and stabilized for 15 min in 800 μl of PBS at 37 °C with 5% CO_2, followed by incubation of the fluorescently labeled HDL in 200 μl of PBS.

1.2.3.5 Method 5: BCθ
L-cell fibroblasts were cultured as described previously in serum-containing media on LabTek two-chambered coverglasses (VWR, Sugarland, TX) overnight [128]. After overnight growth, the medium was removed, washed with PBS, and further incubated with PBS containing BCθ (10 μg ml^{-1}) and 1 mg ml^{-1} fatty-acid-free bovine serum albumin (BSA) at room temperature for 20 min. The cells were washed three times with PBS and then incubated with FITC-Avidin 40 μg ml^{-1} in PBS with BSA at room temperature for another 20 min. The cells were again washed three times with PBS and imaged in PBS.

1.2.4
Confocal Laser Scanning Microscopy (CLSM) and Multiphoton Laser Scanning Microscopy (MPLSM) of Sterol Probes

Synthetic (dansyl-, NBD-, BODIPY-, FCBP), naturally-occurring (DHE), or BCθ probes of cholesterol were incorporated as described above in order to monitor the cholesterol microenvironments within membranes of living L-cells. These fluorescent sterols or sterol probes were imaged in real time by confocal laser scanning microscopy (CLSM) or multiphoton laser scanning microscopy (MPLSM) using a MRC-1024MP Confocal/Multiphoton imaging system (Zeiss, Thornwood, NY), coupled to a Zeiss Axiovert 135 inverted microscope (Zeiss, Thornwood, New York) using either of these Zeiss oil immersion objectives: 40X Apochromat 1.4 NA, a 63× Plan-Apochromat 1.4 NA, or a 100× Fluar 1.3 NA. For CLSM of NBD- and BODIPY-cholesterols the intrinsic excitation system consisted of an Ar^+/Kr^+ ion laser providing a laser line of 488, 568, and 647 nm lines of 3–4 mW at the fiber output. For excitation of the dansyl fluorophore, the system was modified with the addition of a 30 mW temperature stabilized 408 nm diode laser (Power Technology, Little Rock, AR) with beam circularization using anamorphic prisms for excitation. An internal HQ530/40 dichroic filter (Chroma Technology, Rockingham, VT) was used in conjunction with fluorophores emitting in the green wavelength region. For MPLSM of DHE, sub-picosecond multiphoton excitation was provided by a mode-locked Coherent Mira 900F (Coherent, Palo Alto, CA) tuned to 900 nm. The 900 nm infrared pulsed laser light was focused in confocal planes within the cell with the z-axis control. The fluorescence emission from the multiphoton excitation volume was collected by the objective and reflected out of the microscope along the excitation path. Inside the external detector system obtained from Dr. Warren Zipfel (Cornell University, Ithaca, NY), the fluorescence emission was separated from the excitation path with a 670UVDCLP dichroic. Emission was split among three dichroics/filter combinations with corresponding photomultiplier tubes. The three dichroics and emission filters permitted selection of different wavelength/bandwidth regions in three channels. The following three emission filters (Chroma Technology, Rockingham, VT) were used to image DHE in living cells: a filter centered at 375 nm with a 50 nm bandwidth for detection of the higher intensity of the monomeric form; a filter centered at 455 nm with a 30 nm bandwidth for detection of the higher intensity of the microcrystalline form; a filter centered at 400 nm with a 100 nm bandwidth for collecting a significant portion of the DHE spectral emission. Ratioing the intensities from the 455 nm channel to that of the 375 nm channel provides information about the proportion of monomeric to crystalline sterol. In the case where there is no accumulation of the crystalline form, the larger bandwidth enhances the sensitivity by integrating the fluorescence intensity over a larger emissive wavelength region. For the FCBP, a BGG 22 filter was used in combination with a 500 DCLP to cover the range 410–490 nm. Typically four images were averaged using a kalman filter in order to reduce the noise.

1.3 Cholesterol Structure and Distribution in Membranes

The structure of cholesterol consists of a cyclopentanoperhydrophenanthrene type center with a 3-hydroxy group, two methyl groups, a C-5,6 double bond, and an aliphatic side chain with methyl groups (Figure 1.2A) with formula weight $C_{27}H_{46}O$. Cholesterol distributes laterally, not only as cholesterol-rich phases, but also as small clusters (10–25 nm) in model and biological membranes [119, 131, 132], although these cholesterol clusters do not represent cholesterol crystals [128, 133]. The function of these clusters relates to the ability of cholesterol to form a hexagonal phase [134], possibly accounting for the very rapid spontaneous transbilayer migration rate of cholesterol ($t_{1/2}$ = seconds to minutes) in most model and plasma membranes [135–141]. Whether lipid rafts consist of enrichment of these small cholesterol clusters remains unknown.

Consistent with the lower exposure of the cholesterol headgroup (OH) to the aqueous interface in tightly-packed, more highly-ordered membrane regions [65], cholesterol-rich membranes such as lipid rafts have increased transbilayer thickness [2, 142]. Structurally, the degree of cholesterol headgroup aqueous exposure at the membrane/aqueous interface is expected to significantly affect cholesterol desorption/uptake between microdomains and extracellular serum lipoproteins or intracellular cholesterol binding proteins. In bulk plasma membrane analysis, cholesterol has been observed to reside in two microenvironments [84, 143–146]: a more aqueous exposure wherein 10–40% has a long fluorescence lifetime with exchangeable sterol and less exposure wherein 60–90% has a short lifetime with very slowly exchangeable sterol.

Though the transbilayer distribution of cholesterol in lipid rafts remains enigmatic, enrichment in the cytofacial leaflet has been shown indirectly, based upon known localization of other lipids (sphingomyelin (SM) and GM1 in the exofacial leaflet) and cross-linking of GM1 with SM, but not in DRMs [147]. It has been shown that 80–90% of cell cholesterol is in plasma membranes [84], generally distributed in the cytofacial leaflets, as observed both in purified plasma membranes and in intact cells [85, 135, 139, 140, 148–150]. At equilibrium, not only the flip-flop rate, but also phospholipid species, acyl chain unsaturation, and cholesterol–protein interactions contribute significantly to determining the transbilayer cholesterol distribution. For example, some model membrane studies suggest that cholesterol binds phospholipids in the order SM > PS > PC > PE [151], while others show that this order is highly dependent on the specific acyl chain composition (saturated vs. unsaturated) [144, 152]. Thus, while SM is known to be enriched in the plasma membrane outer leaflet [153], whether cholesterol preferentially interacts with SM therein is highly dependent on the acyl chain composition, not only of SM but also of the other phospholipids present in the cytofacial as well as the exofacial leaflets. Enrichment of plasma membrane phospholipids with polyunsaturated fatty acids (PUFAs) esterified to classes other than SM significantly alters transbilayer cholesterol distribution [154, 155]. Cholesterol also preferentially binds with specific membrane proteins. These findings are consistent with the crucial element in cholesterol distribution being the formation of highly packed lipid structures, rather than specific complex formation between SM and cholesterol [144, 156].

Cholesterol efflux from model membranes [84, 143, 145, 146, 157, 158] and plasma membranes [84, 85] has been shown to have two kinetically and structurally resolvable components ($^1t_{1/2}$ = minutes to hours, $^2t_{1/2}$ = days). Since model and plasma membrane cholesterol transbilayer migration rates are rapid (seconds to minutes) [135, 139, 140, 159–161], these cholesterol domains represented lateral, rather than transbilayer, lipid rafts [84, 85, 162, 163]. Slower cholesterol efflux from less fluid model membranes would suggest that rapidly transferable cholesterol must be associated with non-rafts in model membranes [40, 84]. Contrary to expectations, however, cholesterol efflux from plasma membrane lipid rafts was much faster than from non-rafts, suggesting that cholesterol in lipid rafts from biological membranes may not be organized like model membrane microdomains.

Clearly, more studies need to be performed in order to understand fully the interactions of cholesterol with its neighboring lipids and proteins in living membranes and several fluorescent sterol probes have been developed to study different aspects of cholesterol and can be applied to imaging its nature in living cells (Figure 1.2B–G).

1.4
NBD-Cholesterol

The NBD group typically has a broad excitation range that can easily be excited at the 488 nm Ar$^+$ ion laser line with relatively high emission intensity over the region 500–600 nm (Table 1.1) depending upon the polarity of the environment of the probe. In aqueous buffers, the NBD group of the cholesterol analog exhibits a fluorescence

Table 1.1 Fluorescent cholesterol analogs used in microscopic imaging of living cells.

Sterol	Excitation Maximum (nm)		Emission Maximum (nm)		Comments
	EtOH	Aqueous	EtOH	Aqueous	
DHE	311, 324, 340	314, 329, 340	354, 371, 390 (VL)	356, 375, 403, 426 (L)	naturally-occurring, fluidity sensitive, prefers lipid rafts microcrystalline appears in aqueous and lysosomes
NBD-chol	334, 467	425–507	531 (M-H)	550–650 (L)	polarity and fluidity sensitive
BODIPY Chol 2	496	572	504 (H)	573 (M-H)	prefers lipid rafts
Dansyl-chol	336	342	517 (M-H)	506 (L-M)	polarity sensitive, prefers lipid rafts
FCBP	(230, 250, 292)[a] (246, 311)[b]		333, 445 (VL)	ND	photoactivatable, Prefers lipid rafts, binds caveolin

ND = not detectable. Emission yields: VL = very low; L = low; M = medium; H = high.
[a] Excitation maxima for emission collected at 333 nm.
[b] Excitation maxima shifts to 246 and 311 nm when collecting emission at 445 nm.

emission maximum at ~545 nm but shifts to ~533 nm in ethanol (Table 1.1). Two moieties of cholesterol labeled in the alkyl side chain are easily obtainable: 22-(N-(7-nitrobenz-2-*oxa*-1,3-diazol-4-yl)amino)-23,24-bisnor-5-cholen-3β-ol (22-NBD-cholesterol, Figure 1.2C) and 25-(N-[(7-nitrobenz-2-*oxa*-1,3-diazol-4-yl)-methyl]amino)-27-norcholesterol (25-NBD-cholesterol, Figure 1.2E).

1.4.1
22-NBD-Cholesterol

22-NBD-cholesterol was easily incorporated into living cells by direct addition from a stock solution (Figure 1.3A) and by using methyl-β-cyclodextrin complexes (Figure 1.3B); as determined by confocal laser scanning microscopy (CLSM). The images reveal very little propensity for this moiety to be visualized within the L-cell plasma membrane – the cellular membrane with the highest cholesterol content (reviewed in [3, 84–86, 162]. In model membranes composed of dimyristoylphosphatidylcholine (DMPC) and cholesterol, 22-NBD cholesterol was shown to prefer the cholesterol-poor l_d phase as opposed to the cholesterol-rich l_o phase. Using dithionite-mediate quenching of the 22-NBD-cholesterol fluorescence at 30 °C in large unilamellar vesicles of 1-palmitoyl-2-oleoyl-*sn*-glycero-3-phosphocholine (POPC-LUV), the NBD group was partially exposed at the water, suggesting either bending of the flexible alkyl side chain (to which NBD is attached) or potentially an inverted configuration when compared to cholesterol [164]. Although these studies suggested that 22-NBD-cholesterol might insert poorly/weakly into lipid bilayers, other studies showed that 22-NBD-cholesterol was bound with high affinity (nM) by intracellular lipid binding proteins (SCP-2, ADRP) and was bound to these proteins with orientation similar to cholesterol [162, 165–169]. These data indicated that 22-NBD-cholesterol might be a more suitable probe for cholesterol uptake/trafficking

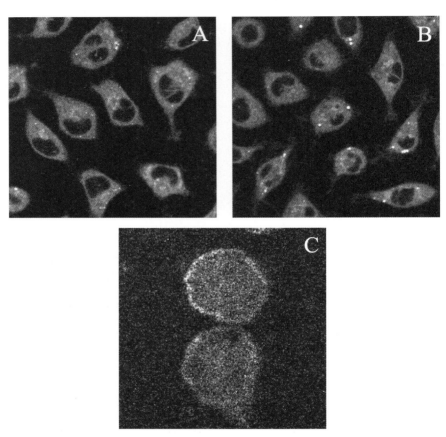

Figure 1.3 CLSM of the 22-NBD-Cholesterol in living cells by three methods. (A) Direct addition: 22-NBD-Chol from a stock solution in EtOH was added to PBS (% EtOH was kept under 0.2% v/v). L-cells, cultured on chambered coverglass, were incubated with the 22-NBD-Chol in PBS so that the incubation concentration was 0.2 µM. Incubation occurred at room temperature for 30 min. (B) 22-NBD-Chol-MβCD complexes: L-cells were incubated with NBD-Chol-MβCD complexes so that the total NBD-cholesterol concentration was 0.6 µM in PBS at room temperature for 40 min. Images were collected as in (A) but with approximately threefold less power. (C) 22-NBD-Chol-HDL complexes: Primary mouse hepatocytes from C57BL/6 mice were incubated in PBS at 37 °C for 30 min with high density lipoproteins loaded with 22-NBD-Cholesterol (0.4 µM) as described in Method 4. All images acquired by CLSM with 488 nm laser excitation and emission detection through a HQ530/40 filter. Zeiss 63× oil immersion objective used in (A) and (B) and the Zeiss 40× oil immersion objective used in (C).

than plasma membrane insertion, at least in cultured L-cell fibroblasts. Consistent with this possibility, 22-NBD-cholesterol cellular uptake rates were >100-fold faster than [^{3}H] cholesterol due to selective uptake via the high density lipoprotein (HDL) receptor mediated pathway [170]. The 22-NBD-cholesterol rapidly enters the living cell and targets lipid droplets through non-vesicular mechanisms [170]. Similarly, efflux of 22-NBD-cholesterol was HDL-mediated with rates much faster than with [^{3}H] cholesterol. Interestingly, although sterol carrier protein-2 (SCP-2) enhanced uptake, it was shown to inhibit the effects of extracellular HDL [169] during efflux, revealing the protein's potential involvement in cellular cholesterol homeostasis [169]. Kinetic analysis of 22-NBD-cholesterol efflux from the cytoplasm of cultured L-cells reveals two dynamic pools: (i) a small (18%), very rapid ($t_{1/2} = 1.9$ min) pool reflecting protein-mediated cholesterol trafficking, and (ii) a large (82%), slower ($t_{1/2} = 14.7$ min) pool reflecting vesicular cholesterol trafficking [169]. Overexpression of sterol carrier protein-2 (SCP-2), which binds 22-NBD-cholesterol with high affinity [162, 165, 166], dramatically accelerated (reduced $t_{1/2}$ to 1.4 min) the rapid, protein-mediated transfer without affecting its pool size. In contrast, SCP-2 overexpression decreased (increased $t_{1/2}$ to 31.4 min) the slower, vesicular transfer – again without affecting its pool size [169]. The latter observation correlated with the ability of SCP-2 to bind and sequester phosphatidylinositides [171, 172] and fatty acyl CoAs [173, 174], both of which are known to regulate vesicular trafficking (reviewed in [175–177]). Incubating primary cultured mouse hepatocytes with high density lipoproteins labeled with 22-NBD-cholesterol, the fluorescent cholesterol analog localizes early into the plasma membrane (Figure 1.3C).

22-NBD-cholesterol has been incorporated into macrophages and lymphocytes obtained from rats. Differences in uptake were examined based upon thioglycollate-injected rats and control rats as well as those stimulated *in vitro* by lipopolysaccharide and phorbol-myristate acetate and compared with concanavalinA treatments which did not appear to modulate cholesterol incorporation by lymphocytes. Thioglycollate-treated macrophages revealed higher initial uptake of NBD-cholesterol [178].

1.4.2
25-NBD-Cholesterol

25-NBD-cholesterol was also easily incorporated into living cells by direct addition from a stock solution (Figure 1.4A) with similar results and no labeling of the plasma membrane even after 30 min (Figure 1.4B–D). However, by using methyl-β-cyclodextrin complexes (Figure 1.5), the plasma membrane was labeled significantly in the first 2 min (Figure 1.5A) but with subsequent diminishment in intensity (Figure 1.5B and C), especially after 10 min as the probe was internalized (Figure 1.5D). The orientation of 25-NBD-cholesterol in membranes is controversial. In model membrane POPC-LUVs, a quenching effect similar to 22-NBD-cholesterol was observed using dithionite quenching, again suggesting either bending of the cholesterol alkyl chain (with attached NBD) toward the bilayer surface or possibly an inverted orientation as compared to cholesterol [164]. However, another study revealed that a second population of sterol in dipalmitoylphosphatidylcholine (DPPC) membranes

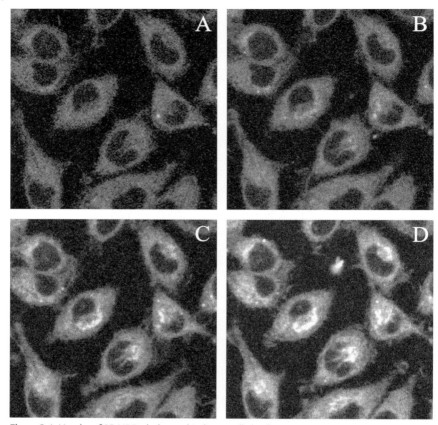

Figure 1.4 Uptake of 25-NBD-cholesterol in living cells by direct addition. 25-NBD-cholesterol from a stock solution in EtOH was added to PBS (% EtOH was kept under 0.2%v/v). L-cells were incubated with 25-NBD-Chol in PBS (incubation concentration of 0.6 μM) at room temperature for (A) 5, (B) 10, (C) 20, (D) 30 min. Images were acquired using CLSM with 488 nm laser excitation and emission detection using a HQ530/40 nm emission filter.

interacted in a tail-to-tail dimer configuration across the membrane bilayer [179]. The intracellular distribution of 25-NBD-cholesterol was reported to be different from that of the 22-NBD-cholesterol with most of the 25-NBD-cholesterol targeting mitochondria [180] in a Chinese hamster ovary (CHO) cell line, as determined by video fluorescence imaging.

In a depth-dependent solvent relaxation study involving the effects of polarity on low amounts of 25-NBD cholesterol within reverse micelles of sodium bis(2-ethylhexyl) sulfosuccinate (AOT) formed in isooctane, although the maximum remained at 513 nm with excitation at 475 nm with increasing amounts of water/AOT ratio, the increasing polarity created an increase in the red edge excitation shift as well as decreases in the intensity maxima and in both components of the fluorophore

1.5 Dansyl-Cholesterol

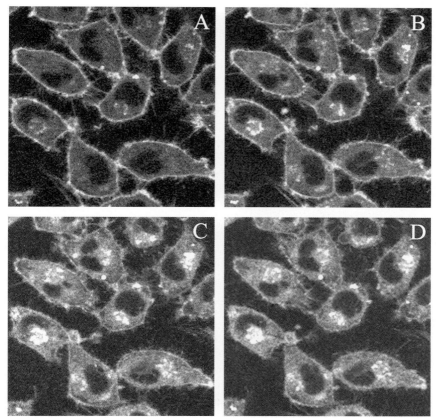

Figure 1.5 Uptake of 25-NBD-cholesterol in living cells using MβCD complexes. L-cells were incubated with 25-NBD-Chol-MβCD (incubation concentration of 25-NBD-cholesterol of 0.6 μM) in PBS at room temperature for (A) 2, (B) 5, (C) 10, (D) 15 min. Images were acquired using CLSM with 488 nm laser excitation and emission detection using a HQ530/40 nm emission filter. CLSM conditions were similar to those in Figure 1.4.

lifetime. With the increase in the water/AOT ratio of 0–25, the red edge excitation shift changed from 6 to 11 nm, reflecting restricted motion as a possible indicator of the fluorophore's location well below the surface [181].

1.5
Dansyl-Cholesterol

Dansyl-cholesterol (DChol) was synthesized recently and used in the study of plasma membrane-derived cholesterol transport involving the Niemann-Pick C1 (NPC1) protein in cultured cells [120]. The structure (Figure 1.2D) of the new probe was that of a cholesterol labeled with the fluorescent dansyl group (5-dimethylamino-1-

naphthalenesulfonyl) attached at carbon position 6 to form the 6-dansyl-cholestanol [120]. In ethanol, the dansyl group has broad excitation with a peak maximum at 336 nm with broad greenish emission peaking at 522 nm (Table 1.1). The emission yield was quite high but photobleaching was significant under high excitation powers. In aqueous buffer with pH 7.4, the DChol emission peak was shifted to 506 nm. Under low pH $\sim$ 1–2, the intensity decreased with a wavelength shift to 490 nm. In lipid bilayers, the dansyl polar group of dansyl lipid probes was shown to reside 19–21 Å from the bilayer center and was sensitive to its environment, creating shifts in the wavelength [182]. CLSM (described above) was performed using a 408 nm laser for excitation and a HQ530/40 nm dichroic emission filter. DChol can be incorporated into living cells by direct addition with incubation periods of 1–2 h (Figure 1.6A), by complexation with methyl-β-cyclodextrin (DChol-MβCD) with short incubation times (Figure 1.6B), and by feeding the cells with LUVs composed of a mixture of POPC:DChol:PS (Figure 1.6C). In all three cases, plasma membrane labeling was evident – especially when the DChol was delivered to the cells as DChol-MβCD complexes or by LUVs.

The DChol probe with the dansyl group located away from the aliphatic side chain and the 3β-OH group (Figure 1.2D) produced different results, unlike the NBD labeled sterols [120]. DChol, in the form of DChol-MBCD complexes, along with [^{3}H] cholesterol was used to label CHO cells and the efflux kinetics, raft association, and esterification rate were measured [120]. The results of the comparisons for uptake and intracellular esterification showed DChol to differ only slightly from [^{3}H] cholesterol in CHO cells [120]. The dansyl cholesterol was reported to traffic to the endoplasmic reticulum with some cells showing accumulation in the Golgi but most going to lipid droplets [120]. The rate of accumulation in lipid droplets seemed to suggest that it was unesterified DChol that appeared at early times [120].

In another study [183] illustrating the different reverse transport pathways of LDL-cholesterol and acetylated LDL-cholesterol, an Olympus IX 70 inverted fluorescence microscope and Imago charge-coupled device was used to observe DChol in fetal liver-derived and bone marrow-derived mouse macrophages. LDL and AcLDL was labeled with DChol and incubated with wild-type and ABCA1-knockout mice-derived macrophages and the results compared to [^{3}H] cholesterol [183]. The LDL-DChol was observed to be transported diffusely through cellular membrane regions whereas the AcLDL-DChol was observed to go to late endosomes [183].

1.6
BODIPY-Cholesterol

Several BODIPY-labeled free cholesterol analogs were synthesized with the hydrophobic BODIPY fluorophore (4,4-difluoro-4-bora-3a,4a-diaza-s-indacene) located in the aliphatic tail [121]. Since the spectral characteristics of the BODIPY include a high coefficient of absorption in the near green with a high quantum yield in the green (Table 1.1) as well as long term photostability, this sterol analog became very

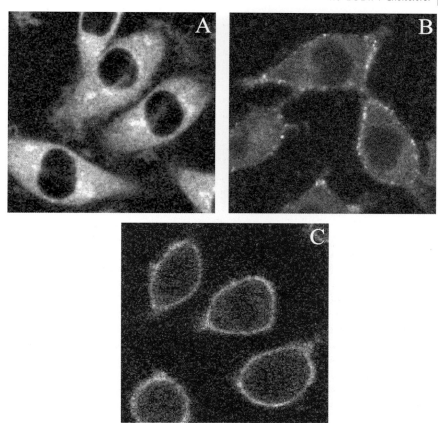

Figure 1.6 The distribution of dansyl-cholesterol within living cells after incubation by three different methods. All images were acquired by CLSM using 408 nm diode laser excitation and a HQ530/40 nm emission filter. (A) Direct addition: Dansyl-cholesterol was added from a stock solution in EtOH to PBS such that the percent EtOH was kept under 0.2%v/v. L-cells were incubated with the Dansyl-cholesterol in PBS with a concentration of 20 μg ml^{-1} at room temperature for 2 h followed by image acquisition. (B) LUV: LUVs containing dansyl-cholesterol (POPC:DChol: DOPS = 55 : 35 : 10 mol%) were made as described in the Methods. L-cells in PBS were incubated with the dansyl cholesterol containing LUVs at 20 μg ml^{-1} at room temperature for 1 h; followed by image acquisition. (C) Dansyl-Chol-MβCD complexes: L-cells were incubated at room temperature with 0.6 μM dansyl-chol-MβCD in PBS for 30 min.

appealing, especially for use in CLSM, as seen in Figure 1.7. Two fluorescent cholesterol analogs, BODIPY-Chol 1, 22-[4-(4,4-difluoro-1,3,5,7-tetramethyl-4-bora-3a,4a,diaza-s-indacen-8-yl)butyroxy]-23,24-bisnorchol-5-en-3β-ol, (not shown) and BODIPY-Chol 2 23-(4,4-difluoro-1,3,5,7-tetramethyl-4-bora-3a,4a,diaza-s-indacen-8-yl)-24-norchol-5-en-3β-ol, (Figure 1.2F) were prepared with the BODIPY linked to the side chain with and without oxygen atoms, respectively, while a third was a BODIPY-coprostanol analog 23-(4,4-difluoro-1,3,5,7-tetramethyl-4-bora-3a,4a,diaza-

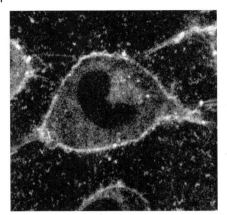

Figure 1.7 Localization of the raft preferring BODIPY-Chol-2 in living cells. LUVs containing BODIPY-Chol-2 (POPC : Chol : BODIPY-Chol-2 : DOPS = 63 : 26 : 7 : 4 mol%) were made as described in the Methods. L-cells in PBS were incubated with the BODIPY-Chol-2 containing LUVs at total sterol concentration 20 µg ml^{-1} at room temperature for 1 h; followed by CLSM image acquisition. Laser excitation was 488 nm with confocal emission detection using the HQ530/40 nm filter.

s-indacen-8-yl)-24-nor-5β-cholestan-3β-ol, (not shown) linked without oxygen atoms [121]. BODIPY-Chol 2 was observed to have ideal mixing in mixed monolayers of the compound with POPC [121]. Further characterization in multilamellar vesicles of the fluorescent sterol and saturated glycerophopholipids or sphingomyelin revealed that BODIPY-Chol 2 compared reasonably well with cholesterol in forming detergent-resistant sterol-rich domains while BODIPY-Chol 1 and 3 did not – suggesting that the BODIPY-Chol 2 might be a preferential probe of cholesterol-rich lipid rafts [121]. The cyclodextrin-mediated desorption rates of cholesterol and BODIPY-Chol 2 from monolayers mixed with POPC and the fluorescent analog were also similar [121].

Further studies using correlated fluorescence-atomic force microscopy revealed that BODIPY-Chol 1 and BODIPY-Chol 3 did not represent cholesterol in supported bilayers [184]. BODIPY-Chol 2 showed similarity with cholesterol but a dependence upon the type of sphingomyelin used: non-preferential distribution between l_o and l_d domains occurred when using N-stearoyl-D-erythro-sphingosylphophorylcholine (18 : 0) but there was preferential distribution into l_o domains using brain sphingomyelin and into l_d domains using the shorter 16 : 0 [184].

Two new BODIPY-labeled analogs of free cholesterol have been synthesized with an acetylenic linkage in the aliphatic portion of the cholesterol through either carbon position 8 or carbon position 5 of the BODIPY fluorophore [185]. In ethanol, the C8 bonded moiety showed an absorbance maximum of 499 nm and an emission maximum of 508 nm while linkage at the BODIPY C5 produced a significant red shift of 70 nm in the absorbance peak and a 75 nm red shift in the emission peak [185]. These compounds also exhibited high extinction coefficients, small Stokes shifts, and high fluorescence yields, characteristic of BODIPY-labeled probes [185].

1.7 Dehydroergosterol (DHE)

A naturally-occurring fluorescent sterol analog, dehydroergosterol (DHE), was found in the membranes of eukaryotes such as the yeast *Candida tropicalis* [186], the yeast *Saccharomyces cerevisiae* [187], and the Red Sea sponge *Biemna fortis* [188]. As a result of its natural occurrence and intrinsic fluorescence, its physical properties have been studied [189] and many investigations undertaken to analyze its characteristics in crystals [122, 124, 190], in model membranes (reviewed in [128, 157, 160, 190–192]), in biological membranes (reviewed in [3, 84–86, 97, 135, 193–196]), and its associations with proteins (reviewed in [162, 193, 197, 198]) including the lipoproteins (reviewed in [199, 200]). Fluorescence resonance energy transfer (FRET) between protein-bound DHE donor and cholesterol binding protein aromatic amino acid acceptor has been used to determine the intermolecular distance between these residues within the sterol binding site [160, 190, 197, 201]. In studies with cultured cells, DHE replaced up to 85% of endogenous cholesterol and had no significant effect on viability of L-cell fibroblasts, CHO cells, macrophages, hepatic cells, or MDCK cells [4, 148, 180, 202–205]. DHE codistributed with cholesterol in the plasma membrane and intracellular membranes and did not affect membrane phospholipid composition, membrane sterol/phospholipid ratio, or function of sterol sensitive enzymes in the plasma membrane [148]. In studies to determine the sterol-structure specificity of cholesterol-sensitive membrane receptors (e.g., oxytocin receptor), when membranes were depleted of cholesterol receptor activity was abolished, but when different sterols were added back, as methyl-β-cyclodextrin complexes, to reconstitute activity only cholesterol and DHE were able to restore nearly all function [206–208]. Finally, as lipid rafts/caveolae have become the subject of intense scrutiny over the past decade DHE has increasingly been employed to examine the static and dynamic properties of sterol in these domains within plasma membranes, both *in vitro* and in living cells [4, 13, 128, 205]. Recent developments in microscopic imaging techniques have enhanced the ability to observe DHE in living cells through conventional or video fluorescence microscopy using ultraviolet excitation [162, 180, 202–204] and by multiphoton laser scanning microscopy (MPLSM) [3, 129, 209, 210].

DHE has a conjugated triene system (Figure 1.2B) which gives the sterol probe an excitation maximum at 324 nm and an emission maximum at 371 nm in ethanol (Table 1.1) but the emission maximum shifts to 404 nm in aqueous buffer [128] because of the formation of microcrystals (Table 1.1). Only a small shift is seen in the absorption or excitation spectrum of the DHE in aqueous buffer, potentially due to the polar environment, but a much larger effect is seen in the emission spectrum. An apparent excimer interaction between DHE molecules within the crystalline structure causes a significant enhancement of the longer wavelength electronic levels [128]. The spectrum reveals a fourfold increase in the quantum efficiency [128] over the quantum efficiency (0.04) measured in ethanol [189]. In solvents such as chloroform or acetone, the quantum yield was reduced dramatically [189]. The ultraviolet excitation of DHE has made imaging by conventional fluorescence microscopy or confocal laser scanning microscopy (CLSM) in living cells difficult due to the potential ultraviolet radiation damage and the need for UV optics, not to mention the high photobleaching rate and low quantum yield [162, 189].

Nevertheless, the fact that DHE is a naturally-occurring fluorescent analog of cholesterol spurred interest such that DHE was eventually imaged successfully in single photon mode at lower excitation intensities using a video imaging approach wherein high ultraviolet (UV) transmissive optics and appropriate filter sets were used in conjunction with a cooled back-thinned charge coupled device (CCD) camera with a high quantum efficiency of detection [180, 202]. Initially, the DHE was delivered to the CHO cells from an ethanol stock solution but difficulties due to aggregation required removal of DHE aggregates by treating the cells with trypsin and extensive washing and centrifugation [180]. In order to facilitate dynamical studies involving DHE intracellular transport, a pulse-chase method utilizing MβCD-DHE complexes was shown to be more effective for quickly loading the plasma membrane with the fluorescent sterol [202–204, 211, 212]. DHE loaded in this manner was shown to colocalize with transferrin in a modified CHO cell line, indicative of DHE distributing into the endocytic recycling compartment (ERC) [211]. Further observations concluded that the DHE transport from the plasma membrane to the ERC was energy independent while transport back to the plasma membrane was vesicular. The results were corroborated using $[^3H]$-cholesterol and ^{125}I-Transferrin isolation techniques [211]. These data suggest that at least some plasma membrane sterol enters the cells by endocytic mechanisms (e.g., clathrin-coated pits) not involving lipid rafts.

In polarized HepG2 human hepatoma cells, DHE delivered to both apical and basolateral membranes was rapidly transferred to the apical membrane through the subapical or apical recycling compartment with very little transfer to the trans-Golgi network [202]. Dehydroergosterol was also incorporated into HDL wherein the apolipoprotein was labeled with the Alexa 488 fluorophore [204]. Using this approach in polarized HegG2 cells, DHE and the Alexa 488-labeled HDL were observed, after 1 min incubation at 37 °C, to accumulate in the bile canaliculus (labeled with rhodamine dextran) within minutes [204]. Interestingly, if the cells underwent depletion of ATP, whereupon uptake of the Alexa 488-HDL and also Alexa 546-Tf (transferrin) was completely blocked, DHE with higher labeling of the basolateral membrane was still observed to accumulate in the bile canaliculus [204], once again indicative of non-vesicular transport. In this study using dehydroergosterol and fluorescence imaging, scavenger receptor class B type I (SR-BI) was observed to mediate the uptake of the HDL-DHE [204]. Other imaging-based techniques, large area fluorescence recovery after photobleaching (FRAP) and fluorescence loss in photobleaching (FLIP) experiments, were performed on DHE in polarized HepG2 cells to study transport in the bile canaliculus and basolateral membrane, revealing a non-vesicular transport pathway [202].

DHE has been incorporated into L-cell fibroblasts where it has been shown to replace nearly 85% of the intracellular cholesterol and maintain cell viability [191]. Incorporation of the probe can be performed using each of the three methods: direct, LUVs, or DHE-MβCD complexes. Direct addition to aqueous buffers creates aggregates of DHE monohydrate crystals which L-cells will phagocytose

1.7 Dehydroergosterol (DHE)

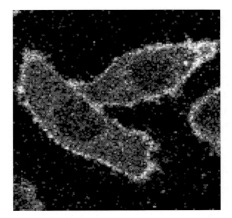

Figure 1.8 Multiphoton laser scanning microscopy of DHE-MβCD in living cells. DHE-MβCD complexes were prepared as described in the Methods. L-cells on chambered cover glass were incubated with the DHE-MβCD complexes (DHE concentration at 20 μg ml^{-1}) in PBS at room temperature for 45 min and imaged. Laser excitation was tuned to 900 nm and the emission was collected using non-descanned detectors and a D400/100 ($\sim$350–450 nm) emission filter.

over time. When imaged by MPLSM, the microcrystals show up brightly due to enhanced longer wavelength emission with a much weaker intensity of the plasma membrane [128]. However, cells can be incubated with LUVs containing DHE which can then fuse to the plasma membrane. DHE taken up this way presents as mostly monomeric under typical conditions [128]. Similarly, cells reveal mostly monomeric emission under uptake by DHE-MβCD complexes (Figure 1.8) with fairly rapid uptake rates, of the order of minutes as opposed to hours. Once it has been verified that monomeric dehydroergosterol has occurred within the living cells, emission filters with broader bandwidth can be used to enhance detection over the emission range of DHE to extend the range of experiments that can be performed.

Previously, few techniques have been applied to directly image the distribution of a suitable cholesterol analog and these did not utilize the optical sectioning ability of multiphoton lasers combined with laser scanning microscopy [213–216]. Now, real-time multiphoton laser scanning microscopy (MPLSM) has been applied to DHE in the plasma membrane of living cells to produce new data and results [124, 128, 129, 162, 209, 217]. MPLSM imaging studies have revealed that there is not a uniform distribution of sterol within the plasma membrane of living murine L-cell fibroblasts [3, 128, 129, 209]. There are regions of high and low concentrations of the DHE, with the DHE-rich regions colocalizing with lipid raft markers. Since the diffraction limit of optical laser microscopy resolution is $\sim$200 nm, the exact nature of the microdomains could not be elucidated. For instance, were the high intensity regions, as represented by several pixels, contiguous lateral sterol domains? In order to further characterize these areas in a more global approach, computerized segmentation techniques along with inference statistics were used to examine the plasma membrane distribution. It was revealed that the sterol-rich regions were distributed non-randomly with evidence of clustering. Estimation of the range of the "macro-level" clustering process was 200–565 nm [209].

1.8
22-(p-Benzoylphenoxy)-23,24-bisnorcholan-5-en-3β-ol (FCBP) Photoactivatable Sterol

This cholesterol analog (Figure 1.2G) with photoactivatable benzophenone group, FCBP, has been synthesized and a radiolabeled version ^{3}H-FCBP prepared [125, 218]. Absorption (Figure 1.9A) and emission spectra (Figure 1.9B) of the FCBP [219] revealed that CLSM was not a realistic possibility due to the UV excitation wavelengths <300 nm and monomeric emission ∼330 nm (Table 1.1) with non-detectable emission in aqueous buffer. On examination of the emission spectrum of the FCBP-MβCD complex (Figure 1.9B), a significant amount of the excimeric emission [220] was observed as well as some shift in emission wavelength (compare Table 1.1). With regards to microscopic imaging in living cells FCBP was a good candidate for multiphoton excitation (MPLSM) but the fluorescence of the monomeric emission in the UV was below the transmission of the microscope optics, including the objective. The excimer emission, however, was shifted into the visible. This was demonstrated by MPLSM imaging as the intracellular fluorescence of the excimer appeared brightly within living cells. The L-cells were labeled by two methods: overnight incubation in the dark with LUVs containing FCBP (Figure 1.9C) or 30 min incubation with FCBP-MβCD complexes (Figure 1.9D).

Previous studies replaced ∼50% of the free cholesterol with the ^{3}H-FCBP in smooth muscle cells during long incubation times (2 days) without dramatic effect upon the cells [125]. Sterol efflux kinetics using apo A-I were similar for ^{3}H-cholesterol and the ^{3}H-FCBP for times up to 5 h. As a result of the stability of the benzophenone group and its ability for UV activation for crosslinking to amino acid α-carbon atoms, further experiments were performed using the smooth muscle cells. These showed that ∼25% of the ^{3}H-FCBP cross-linked to caveolin-1 [125]. The ^{3}H-FCBP was also found in other molecular weight fractions presumably cross-linked to other unidentified proteins in the range 14–150 kDa [125]. In a comparison, using ^{3}H-cholesterol and ^{3}H-FCBP, sterol associated with caveolin-1 decreased in 4–5 h incubation with apolipoprotein A-I (apo A-I) which corresponded to increases in sterol associated with the apo A-I [125].

1.9
BCθ

BCθ, though not a fluorescent sterol itself, has been found to bind cholesterol and so has been found useful in imaging cholesterol distributions in the plasma membrane (Figure 1.10). BCθ was derived by biotinylation of the protease-nicked θ-toxin [126, 127], a cytolysin of *Clostridium perfringens* [126, 127, 221–224]. θ-toxin, or perfringolysin O, has been categorized within a group of thiol-activated hemo-

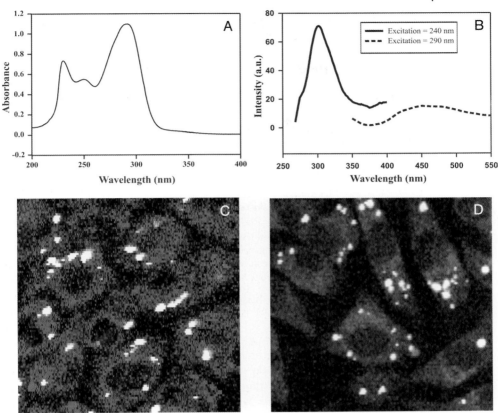

Figure 1.9 FCBP photoactivatable sterol spectral characterization and intracellular excimeric detection by MPLSM. (A) Absorption spectrum of the FCBP in ethanol acquired using a Cary 100 UV–Vis spectrometer. (B) Emission spectra of FCBP-MβCD in PBS at pH 7.2 acquired using a Cary Eclipse spectrofluorometer; solid line showing the monomeric emission obtained with excitation at 240 nm and dashed line showing the excimer emission obtained with excitation at 290 nm. (C) L-cells were incubated with FCBP in 65 : 35 : 10 (mol% POPC : mol% FCPB : mol% PS) LUVs (20 μg ml^{-1} of FCBP) overnight at 37 °C and 5% CO_2. (D) L-cells were incubated with FCBP-MβCD (20 μg ml^{-1} of FCBP) in PBS at room temperature for 30 min. Both images were acquired using 900 nm laser excitation and non-descanned emission detection using a 500DCLP dichroic and BGG22 nm emission filter (~410–490 nm) in combination with the MRC-1024MP. Low intensity sections of the cell were produced by autofluorescence which allowed for cell delineation.

lysins that are inhibited by cholesterol in low quantity and observed to damage the membranes of different cell types by porating the cellular membrane [127, 225]. Perfringolysin O binds cholesterol with $K_d \sim $ nM with the potential to trigger membrane insertion through displacement of a Trp-rich region, as proposed in a recent crystal structure study [225]. From modeling of the ligand with the toxin, the OH of the sterol was thought to form hydrogen bonds with Glu-407 and Arg-457

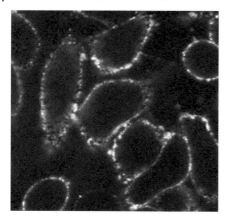

Figure 1.10 Labeling of cholesterol-enriched domains within the plasma membrane using the non-perturbing BCθ. L-cells were incubated with BCθ 10 μg ml^{-1} in PBS containing 1 mg ml^{-1} fatty acid free BSA at room temperature for 20 min. Cells were washed with 0.5 ml of PBS three times and incubated with FITC-Avidin 40 μg ml^{-1} in PBS/BSA at room temperature for 20 min. Again the cells were washed with 0.5 ml of PBS three times and subsequently imaged in PBS at room temperature using CLSM as described in Methods. Laser excitation at 488 nm was used along with confocal detection of FITC emission using a HQ530/40 emission filter.

with the 29 Å long sheet of Domain 4 as the membrane insertion domain, since it could extend throughout much of the membrane bilayer. Cholesterol-rich regions become the source of attraction of the toxin monomers for oligomer development and thus membrane pore formation. Analysis of the C-terminal domain of the toxin revealed that it was essential for targeting and binding to cholesterol-rich domains [226].

It has been shown that the protease-nicked θ-toxin Cθ binds the same sites as the θ-toxin but does not develop the ring structures or pore formation when used at lower temperatures (20 °C) for times less than 20 h [127]. This was significant in that Cθ could be used as a relatively low membrane perturbing probe as compared to filipin which binds cholesterol but will extract it from the membrane [127].

Using the biotinylated form of the Cθ probe, BCθ, both fixed and living Pam 212 cells were treated and labeled with fluorescein-avidin D [227]. The cholesterol distributions, as detected by the BCθ, were imaged by fluorescence microscopy. The mouse transformed keratinocyte Pam 212 cells revealed that BCθ localized in the plasma membrane but did not accumulate in caveolae for cells treated upon ice [227]. After raising the temperature to 37 °C, the fluorescently labeled BCθ was observed in caveolae. The plasma membrane showed strong labeling [227]. BCθ was found to bind selectively to the raft domains within the plasma membrane of platelets isolated from fresh blood [228]. Using sucrose gradients and Triton X-100, BCθ was found in floating low density fractions while treatment with cyclodextrin to remove a third of the cholesterol resulted in a null indication of BCθ in the same fractions [228]. A recent study of Jurkat T-cells employed avidin-magnetic beads to isolate fractions containing bound BCθ and discovered the bound fractions composition to be consistent with lipid rafts. The raft markers, flotillin and GM1 ganglioside, was localized within the BCθ binding fractions [229, 230].

In order to visualize and monitor intracellular trafficking of cholesterol-rich regions such as lipid rafts in living cells, enhanced green fluorescent protein (EGFP) was fused to the N terminus of the domain 4 of the BCθ [226, 230]. This EGFP-D4 was found to clearly label live cells on the plasma membrane with a dependence on cholesterol, as determined by treatment with 2-hydroxypropyl-β-cyclodextrin [226, 230].

In L-cells, BCθ clusters labeled by FITC-Avidin were observed within the plasma membrane (Figure 1.10) with distributions similar to fluorescent cholesterol analogs that have been found in the plasma membrane rafts of intact cells. Due to the variety of stable and highly fluorescent labels for streptavidin (e.g., Alexa Fluors) different wavelengths of excitation and emission can be used when performing colocalization or FRET experiments. Similarly, the use of different green fluorescent protein variants would permit a varied selection of fluorescent wavelengths involving recombinant domain 4 in colocalization and FRET experiments for real-time monitoring of lipid raft dynamics.

1.10
Conclusion

While the fluorescent sterol probes reviewed herein are not an exhaustive list, they nevertheless represent the diversity of approaches for using fluorescence to tag or track the behavior of cholesterol in the cell. In the past, fluorescent sterols have been used primarily to examine the structure, properties, and microenvironment sensed by cholesterol in model membranes, biological membranes (especially plasma membranes), and plasma membrane lipid raft fractions isolated from cells. However, biochemical fractionation may induce perturbations due to the isolation processes used (e.g., detergent, high pH carbonate, etc.). Thus, it is not completely clear if such studies completely reflect the dynamics of cholesterol homeostasis within living cells, either on a global cellular scale or on a microdomain scale. In contrast, non-invasive confocal (CLSM) and multiphoton laser scanning microscopy (MPLSM), though limited in resolution to that of optical microscope objectives (i.e., about 200 nm), provide many unique insights into cholesterol distribution in membranes and cholesterol intracellular trafficking through the use of fluorescent cholesterol analogs and non-perturbing fragments of toxins that bind cholesterol-rich membranes and regions in living cells. As shown by the examples presented herein, the method of introducing fluorescent sterols into the cell (serum, LUV, methyl-β-cyclodextrin, HDL) significantly affects the intracellular distribution and targeting of the fluorescent sterol. Insertion and accumulation of the fluorescent sterol in the plasma membrane is favored by supplementing cells with the fluorescent sterol as methyl-β-cyclodextrin, LUV, or HDL complexes. In addition, the structure of the fluorescent sterol itself is a key determinant of preferential insertion into select plasma membrane microdomains. For example, DHE, DChol, BODIPY-Cholesterol and, to a lesser extent, NBD-cholesterol insert readily into the plasma membrane wherein they appear initially to localize preferentially to lipid raft rather than non-raft domains. In contrast, fluorescent sterol targeting after direct addition to serum-containing medium is highly dependent on the specific fluorescent sterol analog used, for example, NBD-cholesterol targets lipid droplets (L-cells) or mitochondria (CHO cells) while DHE initially accumulates in lysosomes and Golgi followed by lipid droplets and plasma membranes. Thus, depending on the specific fluorescent sterol structure and method of cell incorporation, these tags may monitor/probe select cholesterol uptake/trafficking pathways. The importance of continuing to develop new fluorescent sterol probes for exploring the behavior of cholesterol with imaging technology should be emphasized as no one probe can encompass every aspect perfectly. Corresponding advances in statistical image analysis for global modeling of the properties of these sterol analogs are initiating new possibilities in understanding the behavior of cholesterol.

Acknowledgements

This work was supported in part by the USPHS, NIH GM31651 (FS, ABK), GM72041 (Project 2, ABK,FS), DK70965(BPA), and Mentored Quantitative Research Career Development Award (K25) DK062812 (AMG).

Abbreviations

DHE	dehydroergosterol
DChol	6-dansyl-cholestanol
22-NBD-cholesterol	22-(*N*-(7-nitrobenz-2-*oxa*-1,3-diazol-4-yl)amino)-23,24-bisnor-5-cholen-3β-ol
25-NBD-cholesterol	25-(*N*-[(7-nitrobenz-2-*oxa*-1,3-diazol-4-yl)-methyl]amino)-27-norcholesterol
BODIPY-cholesterol-2	23-(4,4-difluoro-1,3,5,7-tetramethyl-4-bora-3a,4a-diaza-*s*-indacen-8-yl)-24-norchol-5-en-3β-ol)
FCBP	free cholesterol benzophenone
BC-θ	biotinylated Cθ–toxin
FITC-Avidin	fluorescein isothiocyanate-avidin
BHT	butylatedhydroxytoluene
APCI	(atmospheric pressure chemical ionization) mass spectroscopy
HPLC	high performance liquid chromatography
DOPS	1,2-dioleoyl-*sn*-glycero-3-phospho-L-serine
POPC	1-palmitoyl-2-oleoylphosphatidylcholine
LUV	large unilamellar vesicles
MβCD	Methyl-β-cyclodextrin
PBS	phosphate buffered saline
PM	plasma membrane

SCP-2	sterol carrier protein-2
L-FABP	liver fatty acid binding protein
SR-B1	scavenger receptor B1
ABC-A1	ATP-binding cassette protein A1
ABC-G5	ATP-binding cassette protein G5
ABC-G8	ATP-binding cassette protein G8
Pgp	P-glycoprotein
HDL	high density lipoprotein
LDL	low density lipoprotein.

References

1. Bretscher, M.S. and Munro, S. (1993) *Science*, **261**, 1280–1281.
2. Sankaram, M.B. and Thompson, T.E. (1991) *Proceedings of the National Academy of Sciences of the United States of America*, **88**, 8689–8690.
3. Schroeder, F., Atshaves, B.P., Gallegos, A.M., McIntosh, A.L., Liu, J.C., Kier, A.B., Huang, H. and Ball, J.M. (2005) Lipid rafts and caveolae organization, in *Advances in Molecular and Cell Biology* (eds P.G. Frank and M.P. Lisanti), Elsevier, Amsterdam.
4. Gallegos, A.M., Storey, S.M., Kier, A.B., Schroeder, F. and Ball, J.M. (2006) *Biochemistry*, **45**, 12100–12116.
5. Atshaves, B.P., McIntosh, A.L., Payne, H.R., Gallegos, A.M., Landrock, K., Maeda, N., Kier, A.B. and Schroeder, F. (2007) *Journal of Lipid Research*, **48**, 2193–2211.
6. Recktenwald, D.J. and McConnell, H.M. (1981) *Biochemistry*, **20**, 4505–4510.
7. Ipsen, J.H., Karlstrom, G., Mouritsen, O.G., Wennerstrom, H. and Zuckerman, M.J. (1987) *Biochimica et Biophysica Acta*, **905**, 162–172.
8. Schroeder, R., London, E. and Brown, D. (1994) *Proceedings of the National Academy of Sciences of the United States of America*, **91**, 12130–12134.
9. Brown, D.A. and London, E. (1998) *Annual Review of Cell and Developmental Biology*, **14**, 111–136.
10. Brown, D.A. and London, E. (2000) *Journal of Biological Chemistry*, **275**, 17221–17224.
11. Sinha, M., Mishra, S. and Joshi, P.G. (2003) *European Biophysics Journal*, **32**, 381–391.
12. Kahya, N., Scherfield, D., Bacia, K., Poolman, B. and Scheille, P. (2003) *Journal of Biological Chemistry*, **278**, 28109–28115.
13. Gallegos, A.M., McIntosh, A.L., Atshaves, B.P. and Schroeder, F. (2004) *Biochemical Journal*, **382**, 451–461.
14. Mukherjee, S. and Maxfield, F.R. (2004) *Annual Review of Cell Biology*, **20**, 839–866.
15. Shahedi, V., Oradd, G. and Lindblom, G. (2006) *Biophysical Journal*, **91**, 2501–2507.
16. Pike, L.J. (2003) *Journal of Lipid Research*, **44**, 655–667.
17. Pike, L.J. (2004) *Biochemical Journal*, **378**, 281–292.
18. Pike, L.J. (2006) *Journal of Lipid Research*, **47**, 1597–1598.
19. Ikonen, E. and Vainio, S. (2005) *Science's STKE*, **pe3**, 1–3.
20. Smart, E.J. and van der Westhuyzen, D.R. (1998) Scavenger receptors, caveolae, caveolin, and cholesterol trafficking, in *Intracellular Cholesterol Trafficking* (eds T.Y. Chang and D.A. Freeman), Kluwer Academic Publishers, Boston.
21. Babitt, J., Trigatti, B., Rigotti, A., Smart, E.J., Anderson, R.G.W., Xu, S. and

Krieger, M. (1997) *Journal of Biological Chemistry*, **272**, 13242–13249.

22 Lavie, Y., Fiucci, G. and Liscovitch, M. (1998) *Journal of Biological Chemistry*, **273**, 3280–3283.

23 Graf, G.A., Connell, P.M., van der Westhuyzen, D.R. and Smart, E.J. (1999) *Journal of Biological Chemistry*, **274**, 12034–12048.

24 Orso, E., Broccardo, C., Kaminski, W.E., Bottcher, A., Liebisch, G., Drobnik, W., Gotz, A., Chambenoit, O., Diederich, W., Langmann, T., Spruss, T., Luciani, M.-F., Rothe, G., Lackner, K.J., Chimini, G. and Schmitz, G. (2000) *Nature Genetics*, **24**, 192–196.

25 Luker, G.D., Pica, C.M., Kumar, A.S., Covey, D.F. and Piwnica-Worms, D. (2000) *Biochemistry*, **39**, 7651–7661.

26 Demeule, M., Jodoin, J., Gingras, D. and Beliveau, R. (2000) *FEBS Letters*, **466**, 219–224.

27 Mendez, A.J., Lin, G., Wade, D.P., Lawn, R.M. and Oram, J.F. (2001) *Journal of Biological Chemistry*, **276**, 3158–3166.

28 Drobnik, W., Borsukova, H., Bottcher, A., Pfeiffer, A., Liebisch, G., Schutz, G.J., Schindler, H. and Schmitz, G. (2002) *Traffic*, **3**, 268–278.

29 Rigotti, A., Miettinen, H. and Kreiger, M. (2003) *Endocrine Reviews*, **23**, 357–383.

30 Yancey, P.G., Bortnick, A.E., Kellner-Weibel, G., de la Llera-Moya, M., Phillips, M.C. and Rothblat, G.H. (2003) *Arteriosclerosis, Thrombosis, and Vascular Biology*, **23**, 712–719.

31 Peng, Y., Akmentin, W., Conneely, M.A., Lund-Katz, S., Phillips, M.C. and Williams, D.L. (2003) *Molecular Biology of the Cell*, **15**, 384–396.

32 Hinrichs, J.W.J., Klappe, K., Hummel, I. and Kok, J.W. (2004) *Journal of Biological Chemistry*, **279**, 5734–5738.

33 Nieland, T.J.F., Chroni, A., FitzGerald, L.M., Maliga, Z., Zannis, V.I., Kirchhausen, T. and Krieger, M. (2004) *Journal of Lipid Research*, **45**, 1256–1265.

34 Parathath, S., Connelly, M.A., Rieger, R.A., Klein, S.M., Abumrad, N.A., de la Llera-Moya, M., Iden, C.R., Rothblat, G.H. and Williams, D.L. (2004) *Journal of Biological Chemistry*, **279**, 41310–41318.

35 Connelly, M.A. and Williams, D.L. (2004) *Current Opinion in Lipidology*, **15**, 287–295.

36 Smart, E.J. (2005) Caveolae and the regulation of cellular cholesterol homeostasis, in *Advances in Molecular and Cell Biology* (eds M.P. Lisanti and P.G. Frank), Elsevier BV, Amsterdam.

37 Everson, W.V. and Smart, E.J. (2005) Caveolae and the regulation of cellular cholesterol homeostasis, in *Caveolae and Lipid Rafts: Roles in Signal Transduction and the Pathogenesis of Human Disease* (eds M.P. Lisanti and P.G. Frank), Elsevier Academic Press, San Diego.

38 Chao, W.T., Tsai, S.-H., Lin, Y.-C., Lin, W.-W. and Yang, V.C. (2005) *Biochemical and Biophysical Research Communications*, **332**, 743–749.

39 Kamau, S.W., Kramer, S.D., Gunthert, M. and Wunderlich-Allenspach, H. (2005) *In vitro Cellular & Developmental Biology*, **41**, 207–216.

40 Duong, M., Collins, H.L., Jin, W., Zanotti, I., Favari, E. and Rothblat, G. (2006) *Arteriosclerosis, Thrombosis, and Vascular Biology*, **26**, 541–547.

41 Jessup, W., Gelissen, I., Gaus, K. and Kritharides, L. (2006) *Current Opinion in Lipidology*, **17**, 247–267.

42 Barlage, S., Boettcher, D., Boettcher, A., Dada, A. and Schmitz, G. (2006) *Cytometry Part A: Journal of the International Society for Analytical Cytology*, **69**, 196–199.

43 Koonen, D.P.Y., Glatz, J.F., Bonen, A. and Luiken, J.J.F.P. (2005) *Biochimica et Biophysica Acta*, **1736**, 163–180.

44 Ehehalt, R., Fullekrug, J., Pohl, J., Ring, A., Herrmann, T. and Stremmel, W. (2006) *Molecular and Cellular Biochemistry*, **284**, 135–140.

45 Ortegren, U., Karlsson, M., Blazic, N., Blomqvist, M., Nysrom, F.H., Gustavsson, J., Fredman, P. and Stralfors,

P. (2007) *European Journal of Biochemistry*, **271**, 2028–2036.
46. Saltiel, A.R. and Pessin, J.E. (2003) *Traffic*, **4**, 711–716.
47. Cohen, A.W., Combs, T.P., Scherer, P.E. and Lisanti, M.P. (2003) *American Journal of Physiology, Endocrinology and Metabolism*, **285**, E1151–E1160.
48. Balbis, A.B.G., Mounier, C. and Posner, B.I. (2004) *Journal of Biological Chemistry*, **279**, 39348–39357.
49. Kumar, A.S., Xiao, Y.-P., Laipis, P.J., Fletcher, B.S. and Frost, S.C. (2004) *American Journal of Physiology, Endocrinology and Metabolism*, **286**, E568–E576.
50. Elmendorf, J.S. (2004) *Molecular Biotechnology*, **27**, 127–138.
51. Vainio, S., Bykov, I., Hermansson, M., Jokitalo, E., Somerharju, P. and Ikonen, E. (2005) *Biochemical Journal*, **391**, 465–472.
52. Ishikawa, Y., Otsu, K. and Oshikawa, J. (2005) *Cell Signal*, **17**, 1175–1182.
53. Matthews, L.C., Taggart, M.J. and Westwood, M. (2005) *Endocrinology*, **146**, 5463–5473.
54. Vainio, S., Heino, S., Mansson, J.E., Fredman, P., Kuismanen, E., Vaarala, O. and Ikonen, E. (2002) *EMBO Reports*, **3**, 95–100.
55. Ortegren, U., Yin, L., Ost, A., Karlsson, H., Nystrom, F.H. and Stralfors, P. (2006) *FEBS Journal*, **273**, 3381–3392.
56. Syme, C.A., Zhang, L. and Bisello, A. (2006) *Molecular Endocrinology*, **20**, 3400–3411.
57. Rauch, M.C., Ocampo, M.E., Bohle, J., Amthauer, R., Yanez, A.J., Rodriguez-Gil, J.E., Slebe, J.C., Reyes, J.G. and Concha, I.I. (2006) *Journal of Cellular Physiology*, **207**, 397–406.
58. Kahya, N. (2006) *Chemistry and Physics of Lipids*, **141**, 158–168.
59. Anderson, R.G.W. (1993) *Proceedings of the National Academy of Sciences of the United States of America*, **90**, 10909–10913.
60. Yamamoto, M., Toya, fY., Schwencke, C., Lisanti, M.P., Myers, M.G. and Iskikawa, Y. (1998) *The Journal of Biological Chemistry*, **273**, 26962–26968.
61. Smart, E.J., Graf, G.A., McNiven, M.A., Sessa, W.C., Engelman, J.A., Scherer, P.E., Okamoto, T. and Lisanti, M.P. (1999) *Molecular and Cellular Biology*, **19**, 7289–7304.
62. Liu, P., Rudick, M. and Anderson, R.G.W. (2002) *Journal of Biological Chemistry*, **277**, 41295–41298.
63. Stralfors, P. (2005) Insulin Signaling and Caveolae, in *Caveolae and Lipid Rafts: Roles in Signal Transduction and Human Disease* (eds M.P. Lisanti and P.G. Frank), Elsevier Academic Press, San Diego.
64. Igarashi, J. (2005) eNOS regulation by sphingosine 1 phosphate and caveolin, in *Caveolae and Lipid Rafts: Roles in Signal Transduction and the Pathogenesis of Human Disease* (eds M.P. Lisanti and P.G. Frank), Elsevier Academic Press, New York.
65. Morris, R., Cox, H., Mombelli, E. and Quinn, P. (2004) Rafts, Little caves and large potholes: How lipid structure interacts with membrane proteins to create functionally diverse membrane environments, in *Subcellular Biochemistry: Membrane Dynamics and Domains* (ed. P. Quinn), Kluwer Academic/Plenum, New York.
66. Parton, R.G. and Simons, K. (1995) *Science*, **269**, 1398–1399.
67. Jacobson, K. and Dietrich, C. (1999) *Trends in Cell Biology*, **9**, 87–91.
68. Pierce, S.K. (2004) *Nature Cell Biology*, **6**, 180–181.
69. Henderson, R.M., Edwardson, J.M., Geisse, N.A. and Saslowsky, D.E. (2004) *News in Physiological Sciences*, **19**, 39–43.
70. Parton, R.G. (2005) *Science*, **28**, 2404–2405.
71. Lay, S.L. and Kurzchalia, T.V. (2005) *Biochimica et Biophysica Acta*, **1746**, 322–333.
72. Hancock, J.F. (2006) *Nature Reviews. Molecular Cell Biology*, **7**, 456–462.
73. Kenworthy, A.K. (2007) *Journal of Laboratory and Clinical Medicine*, **53**, 312–317.

74 Jacobson, K., Mouritsen, O.G. and Anderson, R.G.W. (2007) *Nature Cell Biology*, **9**, 7–14.
75 Parton, R.G. and Simons, K. (2007) *Nature Reviews. Molecular Cell Biology*, **8**, 185–194.
76 Anderson, R.G.W. and Jacobson, K. (2002) *Science*, **296**, 1821–1825.
77 Kahya, N., Scherfield, D., Bacia, K. and Schwille, P. (2004) *Journal of Structural Biology*, **147**, 77–89.
78 Lommerse, P.H.M., Blab, G.A., Cognet, L., Harms, G.S., Snaar-Jagalska, E., Spaink, H.P. and Schmidt, T. (2004) *Biophysical Journal*, **86**, 609–616.
79 Wisher, M.H. and Evans, W.H. (1975) *Biochemical Journal*, **146**, 375–388.
80 Houslay, M.D. and Stanley, K.K. (1982) *Dynamics of Biological Membranes*, John Wiley and Sons Ltd., New York, NY.
81 Kremmer, T., Wisher, M.H. and Evans, W.H. (1976) *Biochimica et Biophysica Acta*, **455**, 655–664.
82 Brasitus, T.A. and Schachter, D. (1984) *Biochimica et Biophysica Acta*, **774**, 138–146.
83 Sweet, W.D. and Schroeder, F. (1988) Lipid domains and enzyme activity, in *Advances in Membrane Fluidity: Lipid Domains and the Relationship to Membrane Function* (eds R.C. Aloia, C.C. Cirtain and L.M. Gordon), Alan R. Liss, Inc., New York, NY.
84 Schroeder, F., Jefferson, J.R., Kier, A.B., Knittell, J., Scallen, T.J., Wood, W.G. and Hapala, I. (1991) *Proceedings of the Society for Experimental Biology and Medicine*, **196**, 235–252.
85 Schroeder, F., Frolov, A.A., Murphy, E.J., Atshaves, B.P., Jefferson, J.R., Pu, L., Wood, W.G., Foxworth, W.B. and Kier, A.B. (1996) *Proceedings of the Society for Experimental Biology and Medicine*, **213**, 150–177.
86 Schroeder, F., Gallegos, A.M., Atshaves, B.P., Storey, S.M., McIntosh, A., Petrescu, A.D., Huang, H., Starodub, O., Chao, H., Yang, H., Frolov, A. and Kier, A.B. (2001) *Experimental Biology and Medicine*, **226**, 873–890.

87 Hansen, G.H., Immerdal, L., Thorsen, E., Niels-Christiansen, L.-L., Nystrom, B.T., Demant, E.J.F. and Danielsen, E.M. (2001) *Journal of Biological Chemistry*, **276**, 32338–32344.
88 Danielsen, E.M. and Hansen, G.H. (2006) *Molecular Membrane Biology*, **23**, 71–79.
89 Mayor, S. and Rao, M. (2004) *Traffic*, **5**, 231–240.
90 Palade, G.E. (1953) *Journal of Applied Physics*, **24**, 1424.
91 Yamada, E. (1955) *Journal of Biophysical and Biochemical Cytology*, **1**, 445–458.
92 Anderson, R. (1998) *Annual Review of Biochemistry*, **67**, 199–225.
93 Stan, R.V. (2005) *Biochimica et Biophysica Acta*, **1746**, 334–348.
94 Smart, E.J., Ying, Y., Donzell, W.C. and Anderson, R.G. (1996) *Journal of Biological Chemistry*, **271**, 29427–29435.
95 Uittenbogaard, A., Ying, Y.S. and Smart, E.J. (1998) *Journal of Biological Chemistry*, **273**, 6525–6532.
96 Smart, E.J., Ying, Y., Mineo, C. and Anderson, R.G.W. (1995) *Proceedings of the National Academy of Sciences of the United States of America*, **92**, 10404–10408.
97 Atshaves, B.P., Gallegos, A., McIntosh, A.L., Kier, A.B. and Schroeder, F. (2003) *Biochemistry*, **42**, 14583–14598.
98 Foster, L.J., de Hoog, C.L. and Mann, M. (2003) *Proceedings of the National Academy of Sciences of the United States of America*, **100**, 5813–5818.
99 Pike, L.J., Han, X., Chung, K.-N. and Gross, R.W. (2002) *Biochemistry*, **41**, 2075–2088.
100 Heerklotz, H. (2002) *Biophysical Journal*, **83**, 2693–2701.
101 Eckert, G.P., Igbavboa, U., Muller, W. and Wood, W.G. (2003) *Brain Research*, **962**, 144–150.
102 Bae, T.-J., Kim, M.-S., Kim, J.-W., Kim, B.-W., Choo, H.-J., Lee, J.-W., Kim, K.-B., Lee, C.S., Kim, J.-H., Chang, S.Y., Kang, C.-Y., Lee, S.-W. and Ko, Y.-G. (2004) *Proteomics*, **4**, 3536–3548.
103 Blonder, J., Terunuma, A., Conrads, T.P., Chan, K.C., Yee, C., Lucas, D.A., Schaefer, C.F., Yu, L.-R., Issaq, H.J., Veenstra, T.D.

and Vogel, J.C. (2004) *Journal of Investigative Dermatology*, **123**, 691–699.
104 Skwarek, M. (2004) *Archivum Immunologiae et Therapiae Experimentalis*, **52**, 427–431.
105 Kusumi, A. and Suzuki, K. (2005) *Biochimica et Biophysica Acta*, **1746**, 234–251.
106 Silvius, J.R. (2005) *Quarterly Reviews of Biophysics*, **38**, 373–383.
107 MacLellan, D.L., Steen, H., Adam, R.M., Garlick, M., Zurakowshi, D., Gygi, S.P., Freeman, M.R. and Solomon, K.R. (2005) *Proteomics*, **5**, 4733–4742.
108 van Rheenen, J., E. Achame, E.M., Janssen, H., Calafat, J. and Jalink, K. (2005) *EMBO Journal*, **24**, 1664–1673.
109 Babiychuk, E.B. and Draeger, A. (2006) *Biochemical Journal*, **397**, 407–416.
110 Wang, X. and Paller, A.S. (2006) *Journal of Investigative Dermatology*, **126**, 951–953.
111 Banfi, C., Brioschi, M., Wait, R., Begum, S., Gianazza, E., Frato, P., Polvani, G., Vitali, E., Parolari, A., Mussoni, L. and Tremoli, E. (2006) *Proteomics*, **6**, 1976–1988.
112 Martosella, J., Zolotarjova, N., Liu, H., Moyer, S.C., Perkins, P.D. and Boyes, B.E. (2006) *Journal of Proteome Research*, **5**, 1301–1312.
113 Le Naour, F., Andre, M., Boucheix, C. and Rubinstein, E. (2006) *Proteomics*, **6**, 6447–6454.
114 Storey, S.M., Gibbons, T.F., Williams, C.V., Parr, R.D., Schroeder, F. and Ball, J.M. (2007) *Journal of Virology*, **81**, 5472–5483.
115 Feul-Lagerstedt, E., Movitz, C., Pelime, S., Dahlgren, C. and Karlsson, A. (2007) *Proteomics*, **7**, 194–205.
116 Gallegos, A.M., Storey, S.M., Atshaves, B.P., Martin, G.G., Kier, A.B., Ball, J.A. and Schroeder, F. (2007) *Biochemistry*, submitted.
117 London, E. (2005) *Biochimica et Biophysica Acta*, **1746**, 203–320.
118 Shaikh, S.R., Dumaual, A.C., Castillo, A., LoCascio, D., Siddiqui, R.A., Stillwell, W. and Wassall, S.R. (2004) *Biophysical Journal*, **87**, 1752–1766.
119 Stillwell, W. and Wassall, S.R. (2003) *Chemistry and Physics of Lipids*, **126**, 1–27.
120 Wiegand, V., Chang, T.-Y., Strauss, J.F., JIII, Fahrenholz, F. and Gimpl, G. (2003) *FASEB Journal*, **17**, 782–784.
121 Li, Z., Mintzer, E. and Bittman, R. (2005) *Journal of Organic Chemistry*, **71**, 1718–1721.
122 Fischer, R.T., Stephenson, F.A., Shafiee, A. and Schroeder, F. (1985) *Journal of Biological Physics*, **13**, 13–24.
123 Ruyle, W.V., Jacob, T.A., Chemerda, J.M., Chamberlin, E.M., Rosenburg, D.W., Sita, G.E., Erickson, R.L., Aliminosa, L.M. and Tishler, M. (1953) *Journal of the American Chemical Society*, **75**, 2604–2609.
124 McIntosh, A.L., Atshaves, B.P., Gallegos, A.M., Huang, H., Kier, A.B. and Schroeder, F. (2007) Microscopic studies of cholesterol trafficking using dehydroergosterol, *Wiley Encyclopedia of Chemical Biology*, John Wiley & Sons, Inc., Hoboken, NJ.
125 Fielding, P.E., Russell, J.S., Spencer, T.A., Hakamata, H., Nagao, K. and Fielding, C.J. (2002) *Biochemistry*, **41**, 4929–4937.
126 Iwamoto, M., Morita, I., Fukuda, M., Murota, S., Ando, S. and Ohno-Iwashita, Y. (1997) *Biochimica et Biophysica Acta*, **1327**, 222–230.
127 Ohno-Iwashita, Y., Iwamoto, M., Mitsui, K., Ando, S. and Nagai, Y. (1988) *European Journal of Biochemistry*, **176**, 95–101.
128 McIntosh, A., Gallegos, A., Atshaves, B.P., Storey, S., Kannoju, D. and Schroeder, F. (2003) *Journal of Biological Chemistry*, **278**, 6384–6403.
129 McIntosh, A.L., Atshaves, B.P., Huang, H., Gallegos, A.M., Kier, A.B., Schroeder, F., Xu, H., Zhang, W. and Liu, S. (2007) Multiphoton laser scanning microscopy and spatial analysis of dehydroergosterol distributions on plasma membrane of living cells, in *Lipid Rafts* (ed, T.J. McIntosh), Humana Press, Totowa, NJ.
130 Reaven, E., Tsai, L. and Azhar, S. (1995) *Journal of Lipid Research*, **36**, 1602–1617.

131 Tulenko, T.N., Chen, M., Mason, P.E. and Mason, R.P. (1998) *Journal of Lipid Research*, **39**, 947–956.

132 Troup, G.M., Tulenko, T.N., Lee, S.P. and Wrenn, S.P. (2003) *Colloids and Surfaces B: Biointerfaces*, **29**, 217–231.

133 Kruth, H.S., Ifrim, I., Chang, J., Addadi, L., Perl-Treves, D. and Zhang, W.-Y. (2001) *Journal of Lipid Research*, **42**, 1492–1500.

134 Hayakawa, E., Naganuma, M., Mukasa, K., Shimozawa, T. and Araiso, T. (1998) *Biophysical Journal*, **74**, 892–898.

135 Schroeder, F. and Nemecz, G. (1990) Transmembrane Cholesterol Distribution, in *Advances in Cholesterol Research* (eds M. Esfahami and J. Swaney), Telford Press, Caldwell, NJ.

136 Steck, T.L., Ye, J. and Lange, Y. (2002) *Biophysical Journal*, **83**, 2118–2125.

137 Lange, A.J., Dolde, J. and Steck, T.L. (2007) *Journal of Biological Chemistry*, **256**, 5321–5323.

138 Muller, P. and Herrmann, A. (2002) *Biophysical Journal*, **82**, 1418–1428.

139 Schroeder, F., Nemecz, G., Wood, W.G., Joiner, C., Morrot, G., Ayraut-Jarrier, M. and Devaux, P.F. (1991) *Biochimica et Biophysica Acta*, **1066**, 183–192.

140 Wood, W.G., Schroeder, F., Hogy, L., Rao, A.M. and Nemecz, G. (1990) *Biochimica et Biophysica Acta*, **1025**, 243–246.

141 Haynes, M.P., Phillips, M.C. and Rothblat, G.H. (2000) *Biochemistry*, **39**, 4508–4517.

142 Sankaram, M.B. and Thompson, T.E. (1990) *Biochemistry*, **29**, 10676–10684.

143 Nemecz, G. and Schroeder, F. (1988) *Biochemistry*, **27**, 7740–7749.

144 Schroeder, F. and Nemecz, G. (1989) *Biochemistry*, **28**, 5992–6000.

145 Schroeder, F., Nemecz, G., Gratton, E., Barenholz, Y. and Thompson, T.E. (1988) *Biophysical Chemistry*, **32**, 57–72.

146 Schroeder, F., Nemecz, G., Barenholz, Y., Gratton, E. and Thompson, T.E. (1988) Cholestatrienol Time Resolved Fluorescence in Phosphatidylcholine Bilayers, in *Time Resolved Laser Spectroscopy in Biochemistry* (eds J.R. Lakowicz, M. Eftink and J. Wampler), SPIE Press.

147 Pitto, M., Brunner, J., Ferraretto, A., Ravasi, D., Palestini, P. and Masserini, M. (2000) *Glycoconjugate Journal*, **17**, 215–222.

148 Hale, J.E. and Schroeder, F. (1982) *European Journal of Biochemistry*, **122**, 649–661.

149 Jefferson, J.R., Slotte, J.P., Nemecz, G., Pastuszyn, A., Scallen, T.J. and Schroeder, F. (1991) *Journal of Biological Chemistry*, **266**, 5486–5496.

150 Incerpi, S., Jefferson, J.R., Wood, W.G., Ball, W.J. and Schroeder, F. (1992) *Archives of Biochemistry and Biophysics*, **298**, 35–42.

151 Niu, L.L. and Litman, B.J. (2002) *Biophysical Journal*, **83**, 3408–3415.

152 Pitman, M.C. and Suits, F. (2004) *Biochemistry*, **43**, 15318–15328.

153 Daleke, D.L. (2003) *Journal of Lipid Research*, **44**, 233–242.

154 Sweet, W.D. and Schroeder, F. (1988) *FEBS Letters*, **229**, 188–192.

155 Schroeder, F. (1988) Use of fluorescence spectroscopy in the assessment of biological membrane properties, in *Advances in Membrane Fluidity: Methods for Studying Membrane Fluidity* (eds R.C. Aloia, C.C. Cirtain and L.M. Gordon), Alan R. Liss, Inc., New York, NY.

156 Filippov, A., Oradd, G. and Lindblom, G. (2006) *Biophysical Journal*, **90**, 2086–2092.

157 Nemecz, G., Fontaine, R.N. and Schroeder, F. (1988) *Biochimica et Biophysica Acta*, **943**, 511–521.

158 Igbavboa, U., Avdulov, N.A., Schroeder, F. and Wood, W.G. (1996) *Journal of Neurochemistry*, **66**, 1717–1725.

159 Leventis, R. and Silvius, J.R. (2001) *Biophysical Journal*, **81**, 2257–2267.

160 Kubelt, J.K., Muller, P., Wustner, D. and Hermann, A. (2002) *Biophysical Journal*, **83**, 1525–1534.

161 Muller, P. and Herrmann, A. (2002) *Biophysical Journal*, **82**, 1418–1428.

162 Schroeder, F., Frolov, A., Schoer, J., Gallegos, A., Atshaves, B.P., Stolowich, N.J., Scott, A.I. and Kier, A.B. (1998) Intracellular sterol binding proteins, cholesterol transport and membrane domains, in *Intracellular Cholesterol Trafficking* (eds T.Y. Chang and D.A. Freeman), Kluwer Academic Publishers, Boston.

163 Gallegos, A.M., Atshaves, B.P., Storey, S., McIntosh, A., Petrescu, A.D. and Schroeder, F. (2001) *Biochemistry*, **40**, 6493–6506.

164 Scheidt, H.A., Muller, P., Herrmann, A. and Huster, D. (2003) *Journal of Biological Chemistry*, **278**, 45563–45569.

165 Stolowich, N.J., Frolov, A., Petrescu, A.D., Scott, A.I., Billheimer, J.T. and Schroeder, F. (1999) *Journal of Biological Chemistry*, **274**, 35425–35433.

166 Schroeder, F., Frolov, A., Starodub, O., Russell, W., Atshaves, B.P., Petrescu, A.D., Huang, H., Gallegos, A., McIntosh, A., Tahotna, D., Russell, D., Billheimer, J.T., Baum, C.L. and Kier, A.B. (2000) *Journal of Biological Chemistry*, **275**, 25547–25555.

167 Avdulov, N.A., Chochina, S.V., Igbavboa, U., Warden, C.H., Schroeder, F. and Wood, W.G. (1999) *Biochimica et Biophysica Acta*, **1437**, 37–45.

168 Serrero, G., Frolov, A., Schroeder, F., Tanaka, K. and Gelhaar, L. (2000) *Biochimica et Biophysica Acta*, **1488**, 245–254.

169 Atshaves, B.P., Starodub, O., McIntosh, A.L., Roths, J.B., Kier, A.B. and Schroeder, F. (2000) *Journal of Biological Chemistry*, **275**, 36852–36861.

170 Frolov, A., Petrescu, A., Atshaves, B.P., So, P.T.C., Gratton, E., Serrero, G. and Schroeder, F. (2000) *Journal of Biological Chemistry*, **275**, 12769–12780.

171 Schroeder, F., Zhou, M., Swaggerty, C.L., Atshaves, B.P., Petrescu, A.D., Storey, S., Martin, G.G., Huang, H., Helmkamp, G.M. and Ball, J.M. (2003) *Biochemistry*, **42**, 3189–3202.

172 Gadella, T.W. and Wirtz, K.W. (1994) *European Journal of Biochemistry*, **220**, 1019–1028.

173 Frolov, A., Cho, T.H., Billheimer, J.T. and Schroeder, F. (1996) *Journal of Biological Chemistry*, **271**, 31878–31884.

174 Dansen, T.B., Westerman, J., Wouters, F., Wanders, R.J., van Hoek, A., Gadella, T.W. and Wirtz, K.W. (1999) *Biochemical Journal*, **339**, 193–199.

175 Pfanner, N., Glick, B.S., Arden, S.R. and Rothman, J.E. (1990) *Journal of Cell Biology*, **110**, 955–961.

176 Pfanner, N., Orci, L., Glick, B.S., Amherdt, M., Arden, S.R., Malhotra, V. and Rothman, J.E. (1989) *Cell*, **59**, 95–102.

177 Schroeder, F., Atshaves, B.P., McIntosh, A.L., Gallegos, A.M., Storey, S.M., Parr, R.D., Jefferson, J.R., Ball, J.M. and Kier, A.B. (2007) *Biochimica et Biophysica Acta*, **1771**, 700–718.

178 Portioli Silva, E.P., Peres, C.M., Mendonca, J.R. and Curi, R. (2004) *Cell Biochemistry and Function*, **22**, 23–28.

179 Mukherjee, S. and Chattopadhyay, A. (2005) *Chemistry and Physics of Lipids*, **134**, 79–84.

180 Mukherjee, S., Zha, X., Tabas, I. and Maxfield, F.R. (1998) *Biophysical Journal*, **75**, 1915–1925.

181 Kelkar, D.A. and Chattopadhyay, A. (2004) *Journal of Physical Chemistry. B*, **108**, 12151–12158.

182 Asuncion-Punzalan, E., Kachel, K. and London, E. (1998) *Biochemistry*, **37**, 4603–4611.

183 Wang, M.-D., Kiss, R.S., Franklin, V., McBride, H.M., Whitman, S.C. and Marcel, Y.L. (2007) *Journal of Lipid Research*, **48**, 633–645.

184 Shaw, J.E., Epand, R.F., Epand, R.M., Li, Z., Bittman, R. and Yip, C.M. (2006) *Biophysical Journal*, **90**, 2170–2178.

185 Li, Z. and Bittman, R. (2007) *Journal of Organic Chemistry*, **72**, 8367–8382.

186 Sica, D., Boniforti, L. and DiGiacomo, G. (1982) *Phytochemistry*, **21**, 234–236.

187 Bocking, T., Barrow, K.D., Netting, A.G., Chilcott, T.C., Coster, H.G.L. and

Hofer, M. (2000) *FEBS Letters*, **267**, 1607–1618.

188 Delseth, C., Kashman, Y. and Djerassi, C. (1979) *Helvetica Chimica Acta*, **62**, 2037–2045.

189 Smutzer, G., Crawford, B.F. and Yeagle, P.L. (1986) *Biochimica et Biophysica Acta*, **862**, 361–371.

190 Loura, L.M.S. and Prieto, M. (1997) *Biophysical Journal*, **72**, 2226–2236.

191 Schroeder, F. (1984) *Progress in Lipid Research*, **23**, 97–113.

192 Ohvo-Rekila, H., Akerlund, B. and Slotte, J.P. (2000) *Chemistry and Physics of Lipids*, **105**, 167–178.

193 Schroeder, F., Gallegos, A.M., Atshaves, B.P., McIntosh, A., Petrescu, A.D., Huang, H., Chao, H., Yang, H., Frolov, A. and Kier, A.B. (2001) *Experimental Biology and Medicine*, **226**, 873–890.

194 Gallegos, A.M., Atshaves, B.P., Storey, S.M., Schoer, J., Kier, A.B. and Schroeder, F. (2002) *Chemistry and Physics of Lipids*, **116**, 19–38.

195 Schroeder, F. and Wood, W.G. (1995) Lateral lipid domains and membrane function, in *Cell Physiology Source Book* (ed. N. Sperelakis), Academic Press, New York, NY.

196 Schroeder, F., Woodford, J.K., Kavecansky, J., Wood, W.G. and Joiner, C. (1995) *Molecular Membrane Biology*, **12**, 113–119.

197 Fischer, R.T., Cowlen, M.S., Dempsey, M.E. and Schroeder, F. (1985) *Biochemistry*, **24**, 3322–3331.

198 Stolowich, N.J., Petrescu, A.D., Huang, H., Martin, G., Scott, A.I. and Schroeder, F. (2002) *Cellular and Molecular Life Sciences: CMLS*, **59**, 193–212.

199 Schroeder, F., Goh, E.H. and Heimberg, M. (1979) *Journal of Biological Chemistry*, **254**, 2456–2463.

200 Bergeron, R.J. and Scott, J. (1982) *Journal of Lipid Research*, **23**, 391–404.

201 Schroeder, F., Butko, P., Nemecz, G. and Scallen, T.J. (1990) *Journal of Biological Chemistry*, **265**, 151–157.

202 Wustner, D., Herrmann, A., Hao, M. and Maxfield, F.R. (2002) *Journal of Biological Chemistry*, **277**, 30325–30336.

203 Wustner, D., Mondal, M., Tabas, I. and Maxfield, F.R. (2005) *Traffic*, **6**, 396–412.

204 Wustner, D., Mondal, M., Huang, A. and Maxfield, F.R. (2004) *Journal of Lipid Research*, **45**, 427–437.

205 Storey, S.M., Gallegos, A.M., Atshaves, B.P., McIntosh, A.L., Martin, G.G., Landrock, K., Kier, A.B., Ball, J.A. and Schroeder, F. (2007) *Biochemistry*, **46**, 13891–13906.

206 Burger, K., Gimpl, G. and Fahrenholz, F. (2000) *Cellular and Molecular Life Sciences*, **57**, 1577–1592.

207 Gimpl, G., Burger, K. and Fahrenholz, F. (1997) *Biochemistry*, **36**, 10959–10974.

208 Gimpl, G. and Fahrenholz, F. (2000) *European Journal of Biochemistry*, **267**, 2483–2497.

209 Zhang, W., McIntosh, A., Xu, H., Wu, D., Gruninger, T., Atshaves, B.P., Liu, J.C.S. and Schroeder, F. (2005) *Biochemistry*, **44**, 2864–2984.

210 McIntosh, A., Atshaves, B.P., Huang, H., Gallegos, A.M., Kier, A.B., Schroeder, F., Xu, H., Zhang, W. and Liu, S. (2006) Multiphoton laser scanning microscopy and spatial analysis of dehydroergosterol distributions on plasma membranes of living cells, in *Lipid Rafts* (ed. T. McIntosh), Humana Press Inc., Totowa, NJ.

211 Hao, M., Lin, S.X., Karylowski, O.J., Wustner, D., McGraw, T.E. and Maxfield, F.R. (2002) *Journal of Biological Chemistry*, **277**, 609–617.

212 Wustner, D. (2005) *Journal of Microscopy*, **220**, 47–64.

213 Denk, W., Strickler, J.H. and Webb, W.W. (1990) *Science*, **2**, 73–76.

214 Williams, R.M., Zipfel, W.R. and Webb, W.W. (2001) *Current Opinion in Chemical Biology*, **5**, 603–608.

215 Maiti, S., Shear, J.B., Williams, R.M., Zipfel, W.R. and Webb, W.W. (1997) *Science*, **275**, 530–532.

216 Williams, R.M., Shear, J.B., Zipfel, W.R.M.S. and Webb, W.W. (1999) *Biophysical Journal*, **76**, 1835–1846.

217 Gallegos, A.M., Atshaves, B.P., McIntosh, A.L., Storey, S.M., Ball, J.M., Kier, A.B. and Schroeder, F. (2008) *Current Analytical Chemistry*, **4**, 1–7.

218 Wang, P. and Spencer, T.A. (2005) *Journal of Labelled Compounds & Radiopharmaceuticals*, **48**, 781–788.

219 Sun, Y.P., Sears, D.F., Jr. and Saltiel, J. (1989) *Journal of the American Chemical Society*, **111**, 706–711.

220 Heldt, J.R., Heldt, J., Jozefowicz, M. and Kaminski, J. (2001) *Journal of Fluorescence*, **11**, 65–73.

221 Iwamoto, M., Nakamura, M., Mitsui, K., Ando, S. and Ohno-Iwashita, Y. (1993) *Biochimica et Biophysica Acta*, **1153**, 89–96.

222 Nakamura, M., Sekino, N., Iwamoto, M. and Ohno-Iwashita, Y. (1995) *Biochemistry*, **34**, 6513–6520.

223 Ohno-Iwashita, Y., Iwamoto, M., Ando, S., Mitsui, K. and Iwashita, S. (1990) *Biochimica et Biophysica Acta*, **1023**, 441–448.

224 Ohno-Iwashita, Y., Iwamoto, M., Mitsui, K., Ando, S. and Iwashita, S. (1991) *Journal of Biochemistry*, **110**, 369–375.

225 Rossjohn, J., Feil, S.C., McKinstry, W.J., Tweten, R.K. and Parker, M.W. (1997) *Cell*, **89**, 685–692.

226 Shimada, Y., Maruya, M., Iwashita, S. and Ohno-Iwashita, Y. (2002) *European Journal of Biochemistry*, **269**, 6195–6203.

227 Fujimoto, T., Hayashi, M., Iwamoto, M. and Ohno-Iwashita, Y. (1997) *Journal of Histochemistry and Cytochemistry*, **45**, 1197–1205.

228 Waheed, A.A., Shimada, Y., Heijnen, H.F.G., Nakamura, M., Inomata, M., Hayashi, M., Iwashita, S., Slot, J.W. and Ohno-Iwashita, Y. (2001) *Proceedings of the National Academy of Sciences of the United States of America*, **98**, 4926–4931.

229 Shimada, Y., Inomata, M., Suzuki, H., Hayashi, M., Waheed, A.A. and Ohno-Iwashita, Y. (2005) *FASEB Journal*, **272**, 5454–5463.

230 Ohno-Iwashita, Y., Shimada, Y., Waheed, A.A., Hayashi, M., Inomata, M., Nakamura, M., Maruya, M. and Iwashita, S. (2004) *Anaerobe*, **10**, 125–134.

2
Lipid Binding Proteins to Study Localization of Phosphoinositides
Guillaume Halet and Patricia Viard

2.1
Introduction: Phosphoinositide Signaling

In eukaryotic cells, most membrane trafficking events and signal transduction cascades involve lipid–protein interactions, which mediate directional vesicle transport, the assembly of multi-protein complexes and the interaction of enzymes with their substrates. Phosphoinositides (PIs), a unique family of membrane lipids, are specialized in the establishment of membrane–protein interactions, and are major regulators of cellular signaling and homeostasis [1–3].

PIs are phosphorylated derivatives of the lipid phosphatidylinositol (PtdIns), a minor component of biological membranes. So-called lipid kinases phosphorylate PtdIns on positions 3, 4 or 5 of its inositol headgroup to generate seven PI species that are all encountered in mammalian cells (Figure 2.1). These phosphorylation reactions are balanced by lipid phosphatases which dephosphorylate PIs and participate actively in regulating the diversity and dynamics of PI pools in cellular membranes [1–4]. In addition, the activity of PI kinases and phosphatases can be modulated by intracellular signaling pathways, so the cell can adjust the PI content of its membranes to generate the response to the stimuli [1, 3, 5].

PIs are found in virtually all membrane compartments of the cell, for example, the plasma membrane, exocytic and endocytic vesicles and organelle membranes [6, 7]. However, their distribution is not homogenous and most PI species exhibit some enrichment in specific membrane compartments, thus participating in defining organelle "identity" [7]. For example, PtdIns(4,5)P_2 and PtdIns(3,4,5)P_3 are found mostly in the plasma membrane, while the Golgi apparatus appears to be the major cellular pool of PtdIns(4)P (Table 2.1).

Despite their relatively low abundance, PIs regulate a plethora of cellular processes (Table 2.1), including membrane trafficking, cytoskeletal reorganisation, cell migration, cell survival and many intracellular signaling pathways [1, 3, 6–9]. A well known paradigm is the hydrolysis of plasma membrane PtdIns(4,5)P_2 by enzymes of the phospholipase C (PLC) family, which generates the Ca^{2+}-releasing second messenger

Probes and Tags to Study Biomolecular Function. Lawrence W. Miller (Ed.)
Copyright © 2008 WILEY-VCH Verlag GmbH & Co. KGaA, Weinheim
ISBN: 978-3-527-31566-6

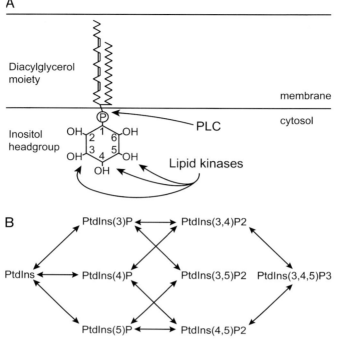

Figure 2.1 (A) Structure of phosphatidylinositol. The sites of action of PLC and lipid kinases are indicated. (B) The mammalian phosphoinositide network. The double arrows indicate phosphorylation/dephosphorylation reactions catalyzed by PI kinases/phosphatases. The generation of PtdIns(5)P via phosphorylation of PtdIns by a specific lipid kinase remains to be demonstrated.

Table 2.1 Phosphoinositide localization and functions in mammalian cells.

PI species	Membrane compartment	Function
PtdIns(3)P	early endosome, phagosome	vesicle trafficking, phagocytosis
PtdIns(4)P	plasma membrane, Golgi	Golgi function, vesicle trafficking, secretion
PtdIns(5)P	nuclear matrix[a]	?
PtdIns(3,4)P_2	plasma membrane	signaling, cell survival
PtdIns(3,5)P_2	late endosomes, lysosomes	vesicle trafficking, protein sorting
PtdIns(4,5)P_2	plasma membrane, Golgi, nuclear matrix[a]	signaling, actin dynamics, vesicle trafficking, RNA processing
PtdIns(3,4,5)P_3	plasma membrane	signaling, actin dynamics, chemotaxis, cell survival

[a]Nuclear PtdIns(5)P and PtdIns(4,5)P_2 are believed to be non-membranous, forming part of proteo-lipid complexes [13–15]. The functions of PtdIns(5)P are still unclear.

inositol 1,4,5-trisphosphate (InsP$_3$) and the lipid diacylglycerol [10]. However, the most fundamental role of PIs is to act as membrane anchors for a variety of cytosolic proteins that possess modular PI-binding domains, thus promoting the assembly of multi-protein complexes, and the activation of effector proteins at specific locations in the cell. For example, the pro-survival serine/threonine kinase Akt interacts with plasma membrane PtdIns(3,4)P$_2$ and PtdIns(3,4,5)P$_3$, resulting in its colocalization with upstream kinases that phosphorylate, hence activate, Akt [8]. Other PI-binding proteins include regulators of small GTPases, cytosolic tyrosine kinases, phospholipases, and proteins regulating membrane trafficking and protein sorting [1, 2, 6–9].

A variety of stimuli induce changes in the PI content of cellular membranes as part of the signal transduction mechanism. Accordingly, dysregulation of PI kinases or phosphatases, or mutations in PI-binding proteins, are linked to many human diseases such as cancer, diabetes, muscular and immune disorders [8, 11]. Understanding PI dynamics and functions has therefore become a major area of research in the past two decades. A range of biochemical techniques are available for detecting PI species and measuring their metabolism in cell extracts [12]. However, these methods are often technically demanding and lack the spatiotemporal resolution necessary for the study of rapid and localized PI signals at the cellular level. The discovery of PI-binding domains, together with major advances in molecular cloning and imaging techniques, have led to the design of PI probes for live cell experiments. These probes, consisting of individual PI-binding domains fused to a fluorescent protein, have dramatically improved our understanding of PI dynamics at the single cell level [6].

In this chapter, we describe the current techniques used to detect PIs in cells, with emphasis on fluorescently-tagged PI-binding domains. Then we show two sets of experiments illustrating the use of PI probes for the detection of PtdIns(3,4,5)P$_3$ in cultured mammalian cells, and PtdIns(4,5)P$_2$ in mouse oocytes.

2.2
Monitoring PI Distribution and Dynamics

2.2.1
Detection of PI Species Using Antibodies

The desire to visualize PIs in cells, with accuracy and specificity, has naturally led to the use of antibodies, for indirect immunofluorescence experiments. Although the generation of such antibodies is technically difficult, since small lipids are poorly immunogenic, several groups have designed immunogen PI derivatives (e.g., liposomal PtdIns(4,5)P$_2$) and have achieved the production of monoclonal antibodies against PIs. Antibodies directed against most PI species are now commercially available, however, most of them are not fully characterized and their use in PI research is still limited.

Several monoclonal antibodies have been raised against PtdIns(4,5)P$_2$ and used in conjunction with fluorescent secondary antibodies to localize PtdIns(4,5)P$_2$

in cells. Interestingly, abundant PtdIns(4,5)P_2 labeling was found in the nuclear matrix of mammalian cells, with a characteristic pattern reminiscent of the nuclear speckles that contain elements of the pre-mRNA splicing machinery [13–16]. Accordingly, nuclear speckles were shown to contain PI kinases [4]. Nuclear PtdIns(4,5)P_2 immunolabeling is resistant to detergent extraction, and therefore must represent a non-membranous pool of PtdIns(4,5)P_2, most likely proteo-lipid complexes [13–15]. Indeed, nuclear PtdIns(4,5)P_2 was found to be complexed with nucleic acids and proteins regulating mRNA processing, and a role for PtdIns(4,5)P_2 in the regulation of pre-mRNA splicing was demonstrated *in vitro* [15].

Monoclonal antibodies were used to investigate changes in PtdIns(4,5)P_2 and PtdIns(3,4,5)P_3 level during cell stimulation. In fibroblasts stimulated with insulin or PDGF, an increase in PtdIns(3,4,5)P_3 immunolabeling was detected in cytoplasmic compartments, while an increase in PtdIns(4,5)P_2 was detected in the nucleus [17]. In human neutrophils stimulated with fMLP, the same antibodies revealed a transient increase in cytosolic PtdIns(3,4,5)P_3 and a concurrent decrease in cytoplasmic PtdIns(4,5)P_2, in agreement with biochemical measurements of PtdIns(4,5)P_2 and PtdIns(3,4,5)P_3 levels after metabolic labeling [17]. However, it is unclear whether the cytoplasmic pool of PtdIns(3,4,5)P_3 is the result of the rapid internalization of plasma membrane PtdIns(3,4,5)P_3, or represents a genuine pool of cytoplasmic PI 3-kinase (PI3K) activity associated with cytoplasmic organelles or proteins.

These studies suggest that antibodies directed against PIs could prove valuable tools in PI research. However, this approach has inherent limitations. First, standard immunolabeling protocols involve membrane permeabilization with detergents, which affect the integrity of the plasma membrane and possibly other cellular membranes. Some relevant PI pools, such as plasma membrane PtdIns(4,5)P_2 and PtdIns(3,4,5)P_3, may therefore be lost during the procedure. Secondly, although antibodies may represent the most specific PI probes, the staining of PIs in fixed cells is of limited relevance when investigating rapid changes in PI level during cell activation. Clearly, the dynamic nature of PI signaling necessitates the use of probes compatible with live cell experimentation.

2.2.2
Fluorescent PI Derivatives

Another strategy for PI visualization in cells is to tag the lipid itself with a fluorophore. Fluorescent derivatives of all PI species have been generated and are commercially available [18]. These synthetic PI analogs are linked to popular fluorophores such as NBD or BODIPY [19]. Their use was facilitated by the development of a new technique for intracellular delivery of synthetic PI in living cells, based on the formation of a complex between the anionic lipid and a cationic polyamine (e.g., neomycin, histone) that acts as a transmembrane carrier [19]. Therefore it has become possible to introduce fluorescent PI into living cells to examine their subcellular localization.

Fluorescent NBD-PtdIns(4,5)P_2 or NBD-PtdIns(3,4,5)P_3 were shuttled into bacteria, yeast, plant cells and various mammalian cell lines [19]. NBD-PtdIns(4,5)P_2 was detected not only in the plasma membrane of mammalian cells, where endogenous PtdIns(4,5)P_2 is found naturally, but also in punctate cytoplasmic structures, and in nuclear speckles, in agreement with antibody labeling experiments [18, 19]. NBD-PtdIns(3,4,5)P_3 localized in the plasma membrane and cytosol but was excluded from the nucleus [19]. Several reports have demonstrated that exogenously added PIs could trigger cellular responses similar to those elicited by endogenous PIs, and serve as a substrate for PI-modifying enzymes [20]. PI shuttling therefore appears as an attractive strategy to study PI localization and functions in living cells. However, the data generated may not be truly representative of the normal metabolism of endogenous PIs. For example, hydrolysis of fluorescent PtdIns(4,5)P_2 by PLC generates fluorescent diacylglycerol in the membrane, which may be mistaken for fluorescent PtdIns(4,5)P_2 when analyzing the images. In addition, the delivery of an extra pool of PI may have adverse consequences on cellular physiology. Thus, intracellular shuttling of PtdIns(4,5)P_2 in fibroblasts triggers an increase in the intracellular Ca^{2+} level, probably as a result of PLC hydrolyzing the extra PtdIns(4,5)P_2 and generating InsP$_3$ [19]. Also, it is unclear at what stage and to what extent exogenous PIs detach from the carrier, and whether the fluorophores affect the properties of the lipids.

2.2.3
Fluorescent PI-Binding Domains

To date, the most attractive method for studying PI distribution and dynamics in living cells is the use of fluorescently-tagged PI-binding domains [6]. These chimeric probes consist of a proteic module, which binds selectively to one PI species, tagged on its C- or N-terminus with a fluorescent protein, such as the green fluorescent protein (GFP) or one of its variants [6, 21]. This technique relies on the simple concept that once expressed into the cell, the GFP-tagged domains will bind to the cellular membranes where their target lipids are localized. Thus, the distribution of individual PI species can be visualized in living cells by fluorescence imaging. These probes are also useful reporters of PI metabolism, since every change in PI content will induce a dynamic redistribution of the fluorescent probes between membrane and cytosol.

Furthermore, since these chimeric constructs are totally proteic, they can be genetically encoded, and expressed into living cells by DNA transfection or RNA injection.

2.2.3.1 Choosing the Right Domain
The choice of the right domain is important. The great majority of PI-binding modules belong to one of several families of conserved protein folds, namely pleckstrin homology (PH), Phox homology (PX), ENTH, FYVE and PHD domains [2, 6]. PI binding involves electrostatic interactions between the negatively charged lipids and positively charged residues in the lipid binding pocket of the protein [2, 6]. The great majority of these domains exhibit little specificity and a low affinity in their binding to PIs, and it is believed that these weak interactions enable signaling proteins to screen

Table 2.2 Commonly used fluorescent PI probes.

PI probe	Parent protein	PI binding	References
2× FYVE$_{Hrs}$-GFP	hepatocyte growth factor-regulated tyrosine kinase substrate	PtdIns(3)P	[25]
GFP-PX$_{p40Phox}$	p40 Phox	PtdIns(3)P	[26–29]
2× FYVE$_{EEA1}$-GFP	early endosome autoantigen 1	PtdIns(3)P	[29, 30]
PH$_{OSBP}$-GFP	oxysterol-binding protein	PtdIns(4)P	[31]
GFP-PH$_{FAPP1}$	PtdIns-four-phosphate adaptor protein 1	PtdIns(4)P	[32]
GFP-PHD$_{ING2}$	inhibitor of growth 2	PtdIns(5)P	[33]
GFP-PH$_{TAPP1}$	tandem PH domain-containing protein 1	PtdIns(3,4)P$_2$	[34, 35]
PH$_{PLC\delta1}$-GFP	phospholipase C delta-1	PtdIns(4,5)P$_2$	[23, 24]
GFP-PH$_{GRP1}$	general receptor for phosphoinositides-1	PtdIns(3,4,5)P$_3$	[36, 37]

PI probes are shown as GFP chimeras, however in some studies GFP variants were used (i.e., Yellow Fluorescent Protein). There is to date no specific fluorescent probe for PtdIns(3,5)P$_2$. 2× indicates the use of two PI binding domains in tandem, which increases the avidity of the interaction.

cellular membranes continuously in search of binding partners or microdomains enriched in PIs. However, some signaling proteins were found to bind selectively to a single PI species, and their PI-binding motifs were used to generate the GFP-tagged PI probes [6]. For example, the PH domain of PLCδ1 (PH$_{PLC\delta1}$) exhibits a strong preference for PtdIns(4,5)P$_2$ [22]. Accordingly, PH$_{PLC\delta1}$ was chosen to generate the PtdIns(4,5)P$_2$ probe, PH$_{PLC\delta1}$-GFP [23, 24]. The most common fluorescent chimeras used to detect PI species in living cells are listed in Table 2.2. It is interesting to note that, to date, all but one PI species can be individually detected using a fluorescent PI-binding domain. A selective PtdIns(3,5)P$_2$-binding motif is yet to be found, but the active research surrounding PtdIns(3,5)P$_2$ signaling, and the recent discovery of PtdIns(3,5)P$_2$ effector proteins, suggest this gap will soon be filled.

2.2.3.2 Imaging PI Probes

GFP-tagged PI probes are visualized in living cells using fluorescence imaging techniques. The most popular technique is confocal microscopy, which generates optical slices of the cell, enabling visualization of fluorescence distribution in the cell interior. PI labeling can thus be easily detected on the inner leaflet of the plasma membrane, or at the surface of cytoplasmic organelles and vesicles. In addition, redistribution of PI probes during the course of an experiment can be followed with time-lapse confocal imaging. Frequently, other fluorescent markers are imaged at the same time as the PI probes, such as plasma membrane markers, actin probes, Ca^{2+} indicators. Therefore, PI metabolism can be examined in conjunction with other cellular processes, such as cell polarization, membrane trafficking and Ca^{2+} signals. Conventional fluorescence microscopy can also be used to image PI probes. In this case, however, the fluorescence signal emanates from the whole cell, resulting in a poor subcellular resolution and substantial background fluorescence. Alternatively, evanescent wave microscopy enables the observation of PI metabolism in the vicinity of the plasma membrane, with little background fluorescence and better spatial resolution than confocal microscopy [38]. This technique requires the microscope to be adapted for evanescent wave excitation.

The following sections show two sets of experiments illustrating PI detection in living cells using GFP-tagged domains and confocal microscopy. The first example investigates agonist-induced *de novo* synthesis of PtdIns(3,4,5)P$_3$, a PI species otherwise absent in quiescent cells. The second example shows the dynamics of a constitutive PI species, plasma membrane PtdIns(4,5)P$_2$, during PLC-induced Ca^{2+} signaling.

2.3
Detection of PtdIns(3,4,5)P$_3$ Synthesis in Transfected Mammalian Cells

PtdIns(3,4,5)P$_3$ is generated from PtdIns(4,5)P$_2$ by PI3K type I, a heterodimer composed of a regulatory protein, such as p85 or p101, and a catalytic subunit p110α, β, δ or γ [8]. This PI is virtually absent in quiescent cells and is generated only after stimulation with growth factors, hormones or cytokines, or upon cell–cell adhesion [6, 8, 36, 37]. PtdIns(3,4,5)P$_3$ recruits and activates a wide range of effector proteins that regulate important cell functions such as metabolism, gene expression, cell growth and survival [8].

In this section, we describe the detection of PtdIns(3,4,5)P$_3$ synthesis in transfected mammalian cells (COS-7) using the probe GFP-PH$_{GRP1}$ (Table 2.2). The PH domain from GRP1 (general receptor for phosphoinositides-1) has been shown to bind specifically to PtdIns(3,4,5)P$_3$ *in vitro* [39] and the translocation of GFP-PH$_{GRP1}$ to the plasma membrane correlates with PtdIns(3,4,5)P$_3$ synthesis, as measured by radioligand displacement assay [37].

2.3.1
Transfection with Plasmid DNA Encoding GFP-PH$_{GRP1}$

The most widely used technique for expressing exogenous proteins in cultured mammalian cells is transfection, the delivery into the cells, by non-viral methods, of DNA or RNA molecules encoding the protein of interest. These genes are generally introduced in the form of plasmid DNA, circular double-stranded DNA molecules that can be replicated by specific bacterial strains for small-scale production and purification. For transfection into mammalian cells, the gene must be placed under the transcriptional control of a promoter recognised by the cellular machinery, such as the human cytomegalovirus immediate early promoter/pCMV (Figure 2.2). Nucleic acids can be introduced into the cells by means of cationic chemical reagents (e.g., calcium phosphate, DEAE-dextran, cationic liposomes) or by a physical method (e.g., microinjection, electroporation). Transfection efficiency varies according to the technique and the cell type. In the following experiments, COS-7 cells were transfected using Fugene6 (Roche Diagnostics Ltd). The cells were plated on glass coverslips and live imaging experiments were performed 2–3 days after transfection. The distribution of GFP-PH$_{GRP1}$ was examined using a Zeiss LSM510 confocal laser

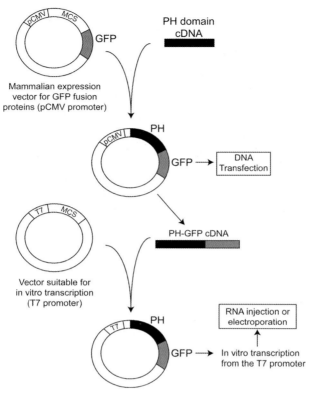

Figure 2.2 How to express PI probes in cells. Illustrated is the construction of a vector encoding a GFP-tagged PH domain. The sequence encoding the PH domain is inserted in the multiple cloning site (MCS) of a mammalian expression vector, in frame with GFP. The resulting fusion protein PH-GFP is expressed in cells by transfection. Alternatively, the sequence encoding PH-GFP can be inserted in a vector containing a T7 promoter, for cRNA synthesis and microinjection.

scanning microscope. Optical slices were 1.5 μm thick. Excitation was provided by the 488 nm line of an argon laser and GFP fluorescence was collected through a bandpass 505–530 emission filter. Confocal images were analyzed with the MetaMorph software (Molecular Devices Corp.).

2.3.2
Detection of PtdIns(3,4,5)P$_3$ Synthesis by a Constitutively-active PI3K

COS-7 cells were transfected with GFP-PH$_{GRP1}$ together with p101 and a modified p110γ that is targeted to the plasma membrane (p110γ-CAAX). As a result, PI3Kγ-CAAX is permanently localized in the vicinity of its substrate PtdIns(4,5)P$_2$ which leads to the constitutive production of PtdIns(3,4,5)P$_3$ [40, 41].

In control cells transfected with GFP-PH$_{GRP1}$ only, the fluorescence of the probe was homogenously distributed in the cytosol, in agreement with PtdIns(3,4,5)P$_3$

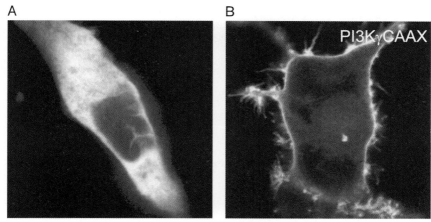

Figure 2.3 PtdIns(3,4,5)P_3 synthesis in COS-7 cells by a constitutively-active PI3K. (A) Control cell expressing the PtdIns(3,4,5)P_3 probe GFP-PH$_{GRP1}$ only. Note the homogenous distribution of the probe (excluding the nucleus). (B) Cell expressing GFP-PH$_{GRP1}$ and PI3Kγ-CAAX. Note the strong accumulation of GFP-PH$_{GRP1}$ at the plasma membrane.

being virtually absent in resting cells. In contrast, in cells also transfected with the constitutively-active PI3Kγ-CAAX, GFP-PH$_{GRP1}$ mostly localized at the plasma membrane (Figure 2.3). This experiment demonstrates that GFP-PH$_{GRP1}$ efficiently detects PI3K activity *in vivo*.

2.3.3
Detection of PtdIns(3,4,5)P_3 Synthesis after Stimulation with EGF

COS-7 cells were transfected with PI3Kα (p85/p110α) and deprived of serum for 24 h prior to experiments. In these conditions, GFP-PH$_{GRP1}$ remained homogenously distributed in the cytosol, indicating that the lipid kinase was not activated (Figure 2.4). After addition of epidermal growth factor (EGF, 250 ng ml^{-1}) to the culture medium, GFP labeling dramatically increased at the plasma membrane whilst cytoplasmic fluorescence decreased, indicating GFP-PH$_{GRP1}$ translocation to the membrane. This translocation was visible as early as 30 s after exposure to EGF, and reached a maximum within approximately 2 min. This experiment illustrates that GFP-PH$_{GRP1}$ can be used as a tool to follow, in real time, the synthesis of PtdIns(3,4,5)P_3 induced by physiological stimuli.

2.4
Monitoring PtdIns(4,5)P_2 Dynamics in Mouse Oocytes

This section illustrates the expression of the PtdIns(4,5)P_2 probe PH$_{PLC\delta1}$-GFP, following injection of the corresponding cRNA in oocytes. Ovulated mouse oocytes

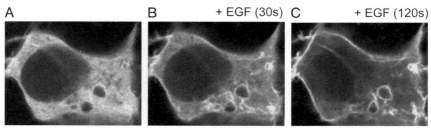

Figure 2.4 PtdIns(3,4,5)P$_3$ synthesis in COS-7 cells stimulated with EGF. Cells were transfected with GFP-PH$_{GRP1}$ and PI3Kα. Before stimulation, PI3Kα is inactive and GFP-PH$_{GRP1}$ remains cytosolic (A). Upon addition of EGF (250 ng ml^{-1}), GFP-PH$_{GRP1}$ translocates to the plasma membrane, indicating PtdIns(3,4,5)P$_3$ synthesis. The translocation was visible at 30 s (B) and reached a maximum at 2 min (C).

are arrested at the metaphase of the second meiotic division (MII), therefore cDNA transfection is irrelevant since transcription is inoperative. This problem is bypassed by injecting polyadenylated cRNAs [42, 43]. In addition, injecting cRNA results in rapid protein expression, since the only limiting step is translation, and imaging can be performed on the day of injection. cRNAs have also been introduced into cultured mammalian cells by electroporation for rapid expression of PI probes [23].

At fertilization, mammalian oocytes undergo a series of Ca^{2+} oscillations which are thought to result from PLC-induced PtdIns(4,5)P$_2$ hydrolysis and InsP$_3$ generation [42, 43]. Therefore, PH$_{PLC\delta1}$-GFP was chosen as a probe to detect changes in the PtdIns(4,5)P$_2$ level in fertilized oocytes. Ca^{2+} oscillations were recorded simultaneously by loading the oocytes with the Ca^{2+} indicator Fura-red.

2.4.1
Making cRNAs

To make cRNA *in vitro* from a cDNA template, the latter must contain a promoter suitable for *in vitro* transcription. Because the common RNA polymerases used for *in vitro* transcription are the phage T7, T3 and SP6 polymerases, the plasmid vector must incorporate the relevant promoter upstream of the cloned coding sequence. We used one of the most popular vectors, pcDNA3.1, which contains a T7 promoter upstream of the multiple cloning site. The sequence encoding PH$_{PLC\delta1}$-GFP was extracted from a mammalian expression vector using restriction enzymes, and inserted in pcDNA3.1 by ligation (Figure 2.2). To generate the DNA template, the plasmid was linearized after the STOP codon using an appropriate restriction enzyme. cRNAs were made *in vitro* using T7 polymerase, and polyadenylated using adequate buffer and reagents that are available from diverse commercial sources [42, 43]. The polyadenylated cRNA was injected in freshly ovulated MII oocytes using pressure-injection. cRNA concentration was adjusted to $\sim$0.2 μg μl^{-1} in the injection pipette and imaging was performed 2–3 h after injection.

2.4.2
PtdIns(4,5)P$_2$ Dynamics in Mouse Oocytes at Fertilization and after Treatment with Ionomycin

Oocytes expressing PH$_{PLC\delta1}$-GFP were freed from their zona pellucida, placed in a glass-bottom chamber heated at 37 °C and scanned using a Zeiss LSM510 confocal microscope. Excitation was provided by the 488 nm line of an argon laser, and GFP and Fura-red fluorescence were collected through a band-pass 505–530 and a long-pass 650 emission filter, respectively. Sperm was added to the chamber and confocal images were acquired at the equator of the oocytes, every 7 s. For post-acquisition analysis, regions of interest were drawn on the confocal time series, using the MetaMorph software, to measure changes in fluorescence intensity. To monitor changes in PH$_{PLC\delta1}$-GFP distribution, plasma membrane (PM) and cytosolic (C) GFP fluorescence were measured individually, and ratioed (PM/C), to cancel artefactual changes in fluorescence intensity that may happen as a result of photobleaching. Fura-red fluorescence was measured in a cytosolic region.

Confocal images revealed that PH$_{PLC\delta1}$-GFP localized exclusively at the plasma membrane of the oocytes, suggesting that the plasma membrane is the main source of PtdIns(4,5)P$_2$ (Figure 2.5A). At fertilization, Ca^{2+} transients were associated with a decrease in cytosolic GFP fluorescence and a corresponding increase in plasma membrane GFP fluorescence, indicating that the probe was recruited at the membrane (Figure 2.5B). This result suggests that there is an increase in plasma membrane PtdIns(4,5)P$_2$ level during Ca^{2+} increase. Further experiments have shown that this transient increase in plasma membrane PtdIns(4,5)P$_2$ is a consequence of cortical granule exocytosis [42]. It is possible that fertilization induced a moderate hydrolysis of plasma membrane PtdIns(4,5)P$_2$ that was masked by the PtdIns(4,5)P$_2$ increase due to exocytosis [42].

In contrast, when unfertilized oocytes were exposed to the Ca^{2+} ionophore ionomycin, which triggers a Ca^{2+} increase that activates endogenous PLCs [24], PtdIns(4,5)P$_2$ was rapidly hydrolyzed, as shown by the rapid redistribution of the probe from the membrane to the cytosol (Figure 2.6). Since this experiment was realized in the absence of extracellular Ca^{2+}, ionomycin acted on the intracellular Ca^{2+} stores, and the resulting Ca^{2+} increase was transient. The return of the Ca^{2+} level to the baseline was associated with the translocation of PH$_{PLC\delta1}$-GFP back to the plasma membrane, demonstrating PtdIns(4,5)P$_2$ resynthesis (Figure 2.6).

2.5
Limitations of the Technique

The GFP-tagged PI probes mentioned in this chapter (Table 2.1) have proven useful tools for research and are used routinely by an increasing number of laboratories. However, the interpretation of the data generated is sometimes complicated by the intrinsic properties of the protein domains. Some specific points on which caution is recommended are listed below.

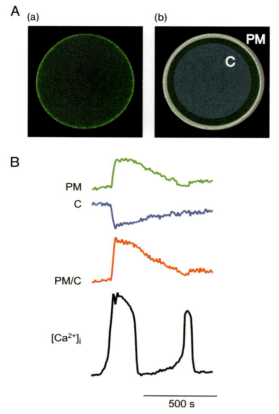

Figure 2.5 Increase in plasma membrane PtdIns(4,5)P_2 in a mouse oocyte during fertilization. The PtdIns(4,5)P_2 probe PH$_{PLC\delta1}$-GFP accumulates at the plasma membrane of the unfertilized oocyte (A(a)). Regions of interest corresponding to the plasma membrane (PM) and the cytosol (C) were drawn on the confocal images (A(b)) to monitor changes in fluorescence intensity. Cytosolic Ca^{2+} was measured simultaneously using the Ca^{2+} indicator Fura-red. Confocal images were acquired at a rate of one frame every 7 s. During the first fertilization Ca^{2+} transient, PM fluorescence increases and cytosolic fluorescence decreases, demonstrating that the probe translocates from the cytosol to the plasma membrane. This increase in plasma membrane PtdIns(4,5)P_2 is illustrated by the ratio PM/C. Adapted from [42].

2.5.1
PI Probes may Detect Only a Subset of the Total PI Pool

Biochemical data have demonstrated the presence of PtdIns(4,5)P_2 in Golgi membranes, however the PtdIns(4,5)P_2 probe PH$_{PLC\delta1}$-GFP was not found to localize to the Golgi. In the same way, GFP-tagged PI probes do not accumulate at nuclear speckles where PIs are thought to be found. Therefore, some PI pools escape detection by GFP-tagged domains. To explain this discrepancy, it was suggested

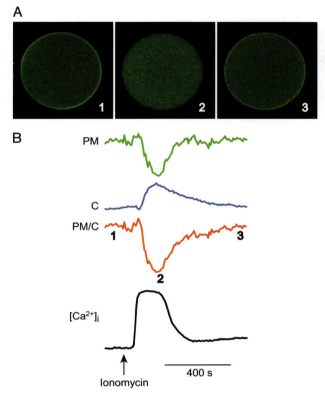

Figure 2.6 PtdIns(4,5)P$_2$ hydrolysis after stimulation of a mouse oocyte with ionomycin. Changes in PtdIns(4,5)P$_2$ level (PM/C) and [Ca^{2+}]$_i$ were monitored as described in Figure 2.5. The transient increase in [Ca^{2+}]$_i$ induced by the ionophore triggered a translocation of PH$_{PLC\delta1}$-GFP from the plasma membrane to the cytosol, reflecting PtdIns(4,5)P$_2$ hydrolysis. Upon return of [Ca^{2+}]$_i$ to the resting level, PH$_{PLC\delta1}$-GFP translocated back to the plasma membrane, demonstrating PtdIns(4,5)P$_2$ resynthesis. Confocal images taken before (1), during (2) and after (3) the Ca^{2+} signal are shown in panel A, and the corresponding time points are indicated on the PM/C trace (B). The final concentration of ionomycin was 1 μM.

that PI probes recognize their target lipid only in a certain molecular context, that is, by interacting also with neighboring lipids or auxiliary proteins [44, 45]. For instance, localization of PH$_{OSBP}$-GFP or GFP-PH$_{FAPP1}$ to the Golgi (Table 2.2) is due to their binding to both PtdIns(4)P and the small GTPase ARF1 at the surface of Golgi membranes [31, 32]. Since this interaction is strictly dependent on the presence of PtdIns(4)P, these probes are still good markers for Golgi PtdIns(4)P, but they cannot detect PtdIns(4)P in the plasma membrane where the lipid is not colocalized with ARF1. Therefore, when analyzing experimental data, it is important to keep in mind that the localization of PI probes in the cell may reveal only a subset of the total PI pool.

2.5.2
PI Probes can Interfere with Normal Cellular Function

Overexpression of PI-binding domains in living cells can result in the sequestration of a significant pool of PIs. Therefore, the fluorescent PI probes may compete with endogenous proteins for binding PIs, and disturb PI-dependent signaling pathways. For this reason, cells expressing an exaggerated amount of the probes (e.g., after transfection) are often discarded from functional studies. However, for the same reason, PI probes can help define the physiological roles of a particular PI species. Thus, when expressed at high levels, GFP-tagged PtdIns(3,4,5)P_3-binding PH domains inhibit Akt activation, cell adhesion and cell spreading, processes which are known to be regulated by PtdIns(3,4,5)P_3 [45]. In the same way, overexpressed GFP-PH$_{FAPP1}$ inhibits vesicle transport from the Golgi to the plasma membrane, as a result of PtdIns(4)P sequestration on the Golgi [32].

2.5.3
Binding of PI Probes to Inositol Phosphates

This is exemplified by PH$_{PLC\delta1}$, which, *in vitro*, exhibits a higher affinity for InsP_3, the inositol headgroup of PtdIns(4,5)P_2, than for PtdIns(4,5)P_2 itself [39, 46]. Therefore, while PH$_{PLC\delta1}$-GFP is a good probe for labeling plasma membrane PtdIns(4,5)P_2 in resting cells, its release from the plasma membrane during PLC activation may reflect competitive binding of InsP_3 to the PH domain, rather than a decrease in the plasma membrane PtdIns(4,5)P_2 level. In fact, many investigators are using PH$_{PLC\delta1}$-GFP as an InsP_3 probe, by monitoring changes in cytosolic fluorescence [6]. However, it is more likely that the dynamics of PH$_{PLC\delta1}$-GFP during PLC activation represent a combination of competitive InsP_3 binding and decrease in PtdIns(4,5)P_2 level [47].

2.6
Conclusion

Since the discovery of the PI cycle, cell research has become increasingly focused on PIs as more PI-dependent cellular processes have been discovered. The importance of PIs as signaling intermediates is emphasized by the fact that disturbing PI signaling can lead to various human diseases. Hence the need to understand PI dynamics at the cellular level and, for this purpose, fluorescent PI probes are certainly the best tools developed to date.

Many new PI-binding proteins are yet to be characterized, and possibly new families of PI-binding domains will be discovered. This could lead to the design of new PI probes, for the detection of PI pools that remain difficult to investigate, such as nuclear PIs or PtdIns(3,5)P_2. In addition, the importance of PI signaling has been demonstrated in a range of organisms, from yeast to humans, and is emerging in plant cells. Therefore, PI signaling appears to be a universal mechanism, just as Ca^{2+}

signaling or protein phosphorylation, and PI probes will certainly be an essential piece of the cell signaling toolkit for the next decades.

Abbreviations

PI	phosphoinositide
PtdIns	phosphatidylinositol
PtdIns(3)P	phosphatidylinositol 3-phosphate
PtdIns(4)P	phosphatidylinositol 4-phosphate
Ptd-Ins(5)P	phosphatidylinositol 5-phosphate
PtdIns(4,5)P_2	phosphatidylinositol 4,5-bisphosphate
PtdIns(3,4)P_2	phosphatidylinositol 3,4-bisphosphate
PtdIns(3,5)P_2	phosphatidylinositol 3,5-bisphosphate
PtdIns(3,4,5)P_3	phosphatidylinositol 3,4,5-trisphosphate
PLC	phospholipase C
InsP_3	inositol 1,4,5-trisphosphate
GFP	green fluorescent protein
PDK1	3-phosphoinositide-dependent kinase-1
PI3K	phosphoinositide 3-kinase
PH	pleckstrin homology

Acknowledgements

The authors wish to thank the Medical Research Council and the Wellcome Trust for funding. P. Viard is the recipient of a Wellcome Trust Research Career Development Fellowship.

References

1 Takenawa, T. and Itoh, T. (2001) Phosphoinositides, key molecules for regulation of actin cytoskeletal organization and membrane traffic from the plasma membrane. *Biochimica et Biophysica Acta*, **1533**, 190–206.
2 Lemmon, M.A. (2003) Phosphoinositide recognition domains. *Traffic*, **4**, 201–213.
3 Balla, T. (2006) Phosphoinositide-derived messengers in endocrine signaling. *Journal of Endocrinology*, **188**, 135–153.
4 Anderson, R.A., Boronenkov, I.V., Doughman, S.D., Kunz, J. and Loijens, J.C. (1999) Phosphatidylinositol phosphate kinases, a multifaceted family of signaling enzymes. *Journal of Biological Chemistry*, **274**, 9907–9910.
5 Payrastre, B., Missy, K., Giuriato, S., Bodin, S., Plantavid, M. and Gratacap, M.P. (2001) Phosphoinositides-Key players in cell signalling, in time and space. *Cell Signal*, **13**, 377–387.
6 Halet, G. (2005) Imaging phosphoinositide dynamics using GFP-tagged protein domains. *Biology of the Cell*, **97**, 501–518.
7 Behnia, R. and Munro, S. (2005) Organelle identity and the signposts

for membrane traffic. *Nature*, **438**, 597–604.

8 Vanhaesebroeck, B., Leevers, S.J., Ahmadi, K., Timms, J., Katso, R., Driscoll, P.C., Woscholski, R., Parker, P.J. and Waterfield, M.D. (2001) Synthesis and function of 3-phosphorylated inositol lipids. *Annual Review of Biochemistry*, **70**, 535–602.

9 Godi, A., Di Campli, A., Konstantakopoulos, A., Di Tullio, G., Alessi, D.R., Kular, G.S., Daniele, T., Marra, P., Lucocq, J.M. and De Matteis, M.A. (2004) FAPPs control Golgi-to-cell-surface membrane traffic by binding to ARF and PtdIns(4)P. *Nature Cell Biology*, **6**, 393–404.

10 Berridge, M.J., Bootman, M.D. and Roderick, H.L. (2003) Calcium signalling: dynamics, homeostasis and remodelling. *Nature Reviews. Molecular Cell Biology*, **4**, 517–529.

11 Pendaries, C., Tronchère, H., Plantavid, M. and Payrastre, B. (2003) Phosphoinositide signaling disorders in human diseases. *FEBS Letters*, **546**, 25–31.

12 Rusten, T.E. and Stenmark, H. (2006) Analyzing phosphoinositides and their interacting proteins. *Nature Methods*, **3**, 251–258.

13 Boronenkov, I.V., Loijens, J.C., Umeda, M. and Anderson, R.A. (1998) Phosphoinositide signaling pathways in nuclei are associated with nuclear speckles containing pre-mRNA processing factors. *Molecular Biology of the Cell*, **9**, 3547–3560.

14 Thomas, C.L., Steel, J., Prestwich, G.D. and Schiavo, G. (1999) Generation of phosphatidylinositol-specific antibodies and their characterization. *Biochemical Society Transactions*, **27**, 648–652.

15 Osborne, S.L., Thomas, C.L., Gschmeissner, S. and Schiavo, G. (2001) Nuclear PtdIns(4,5)P_2 assembles in a mitotically regulated particle involved in pre-mRNA splicing. *Journal of Cell Science*, **114**, 2501–2511.

16 McVey Ward, D., Shiflett, S.L., Huynh, D., Vaughn, M.B., Prestwich, G. and Kaplan, J. (2003) Use of expression constructs to dissect the functional domains of the CHS/beige protein: identification of multiple phenotypes. *Traffic*, **4**, 403–415.

17 Chen, R., Kang, V.H., Chen, J., Shope, J.C., Torabinejad, J., DeWald, D.B. and Prestwich, G.D. (2002) A monoclonal antibody to visualize PtdIns(3,4,5)P_3 in cells. *Journal of Histochemistry and Cytochemistry*, **50**, 697–708.

18 Prestwich, G.D. (2004) Phosphoinositide signaling: from affinity probes to pharmaceutical targets. *Chemistry & Biology*, **11**, 619–637.

19 Ozaki, S., DeWald, D.B., Shope, J.C., Chen, J. and Prestwich, G.D. (2000) Intracellular delivery of phosphoinositides and inositol phosphates using polyamine carriers. *Proceedings of the National Academy of Sciences of the United States of America*, **97**, 11286–11291.

20 Prestwich, G.D. (2005) Visualization and perturbation of phosphoinositide and phospholipid signaling. *Prostaglandins & Other Lipid Mediators*, **77**, 168–178.

21 Lippincott-Schwartz, J. and Patterson, G.H. (2003) Development and use of fluorescent protein markers in living cells. *Science*, **300**, 87–91.

22 Lemmon, M.A., Ferguson, K.M., O'Brien, R., Sigler, P.B. and Schlessinger, J. (1995) Specific and high-affinity binding of inositol phosphates to an isolated pleckstrin homology domain. *Proceedings of the National Academy of Sciences of the United States of America*, **92**, 10472–10476.

23 Stauffer, T.P., Ahn, S. and Meyer, T. (1998) Receptor-induced transient reduction in plasma membrane PtdIns(4,5)P_2 concentration monitored in living cells. *Current Biology*, **8**, 343–346.

24 Várnai, P. and Balla, T. (1998) Visualization of phosphoinositides that bind pleckstrin homology domains: calcium- and agonist-induced dynamic changes and relationship to myo-[^{3}H]inositol-labeled phosphoinositide pools. *Journal of Cell Biology*, **143**, 501–510.

25. Gillooly, D.J., Morrow, I.C., Lindsay, M., Gould, R., Bryant, N.J., Gaullier, J.M., Parton, R.G. and Stenmark, H. (2000) Localization of phosphatidylinositol 3-phosphate in yeast and mammalian cells. *EMBO Journal*, **19**, 4577–4588.
26. Ellson, C.D., Anderson, K.E., Morgan, G., Chilvers, E.R., Lipp, P., Stephens, L.R. and Hawkins, P.T. (2001a) Phosphatidylinositol 3-phosphate is generated in phagosomal membranes. *Current Biology*, **11**, 1631–1635.
27. Ellson, C.D., Gobert-Gosse, S., Anderson, K.E., Davidson, K., Erdjument-Bromage, H., Tempst, P., Thuring, J.W., Cooper, M.A., Lim, Z.Y., Holmes, A.B., Gaffney, P.R.J., Coadwell, J., Chilvers, E.R., Hawkins, P.T. and Stephens, L.R. (2001b) PtdIns(3)P regulates the neutrophil oxidase complex by binding to the PX domain of p40^{phox}. *Nature Cell Biology*, **3**, 679–682.
28. Kanai, F., Liu, H., Field, S.J., Akbary, H., Matsuo, T., Brown, G.E., Cantley, L.C. and Yaffe, M.B. (2001) The PX domains of p47phox and p40phox bind to lipid products of PI(3)K. *Nature Cell Biology*, **3**, 675–678.
29. Scott, C.C., Cuellar-Mata, P., Matsuo, T., Davidson, H.W. and Grinstein, S. (2002) Role of 3-phosphoinositides in the maturation of Salmonella-containing vacuoles within host cells. *Journal of Biological Chemistry*, **277**, 12770–12776.
30. Vieira, O.V., Botelho, R.J., Rameh, L., Brachmann, S.M., Matsuo, T., Davidson, H.W., Schreiber, A., Backer, J.M., Cantley, L.C. and Grinstein, S. (2001) Distinct roles of class I and class III phosphatidylinositol 3-kinases in phagosome formation and maturation. *Journal of Cell Biology*, **155**, 19–25.
31. Levine, T.P. and Munro, S. (2002) Targeting of Golgi-specific pleckstrin homology domains involves both PtdIns 4-kinase-dependent and -independent components. *Current Biology*, **12**, 695–704.
32. Godi, A., Di Campli, A., Konstantakopoulos, A., Di Tullio, G., Alessi, D.R., Kular, G.S., Daniele, T., Marra, P., Lucocq, J.M. and De Matteis, M.A. (2004) FAPPs control Golgi-to-cell-surface membrane traffic by binding to ARF and PtdIns(4)P. *Nature Cell Biology*, **6**, 393–404.
33. Gozani, O., Karuman, P., Jones, D.R., Ivanov, D., Cha, J., Lugovskoy, A.A., Baird, C.L., Zhu, H., Field, S.J., Lessnick, S.L., Villasenor, J., Mehrotra, B., Chen, J., Rao, V.R., Brugge, J.S., Ferguson, C.G., Payrastre, B., Myszka, D.G., Cantley, L.C., Wagner, G., Divecha, N., Prestwich, G.D. and Yuan, J. (2003) The PHD finger of the chromatin-associated protein ING2 functions as a nuclear phosphoinositide receptor. *Cell*, **114**, 99–111.
34. Kimber, W.A., Trinkle-Mulcahy, L., Cheung, P.C.F., Deak, M., Marsden, L.J., Kieloch, A., Watt, S., Javier, R.T., Gray, A., Downes, C.P., Lucocq, J.M. and Alessi, D.R. (2002) Evidence that the tandem-pleckstrin-homology-domain-containing protein TAPP1 interacts with Ptd(3,4)P$_2$ and the multi-PDZ-domain-containing protein MUPP1 *in vivo*. *Biochemical Journal*, **361**, 525–536.
35. Marshall, A.J., Krahn, A.K., Ma, K., Duronio, V. and Hou, S. (2002) TAPP1 and TAPP2 are targets of phosphatidylinositol 3-kinase signaling in B cells: sustained plasma membrane recruitment triggered by the B-cell antigen receptor. *Molecular and Cellular Biology*, **22**, 5479–5491.
36. Venkateswarlu, K., Gunn-Moore, F., Oatey, P.B., Tavaré, J.M. and Cullen, P.J. (1998) Nerve growth factor- and epidermal growth factor-stimulated translocation of the ADP-ribosylation factor-exchange factor GRP1 to the plasma membrane of PC12 cells requires activation of phosphatidylinositol 3-kinase and the GRP1 pleckstrin homology domain. *Biochemical Journal*, **335**, 139–146.
37. Gray, A., van der Kaay, J. and Downes, C.P. (1999) The pleckstrin homology domains of protein kinase B and GRP1 (general receptor for phosphoinositides-1) are sensitive and selective probes for the

38 Haugh, J.M., Codazzi, F., Teruel, M. and Meyer, T. (2000) Spatial sensing in fibroblasts mediated by 3′ phosphoinositides. *Journal of Cell Biology*, **151**, 1269–1279.

cellular detection of phosphatidylinositol 3,4-bisphosphate and/or phosphatidylinositol 3,4,5-trisphosphate in vivo. *Biochemical Journal*, **344**, 929–936.

39 Kavran, J.M., Klein, D.E., Lee, A., Falasca, M., Isakoff, S.J., Skolnik, E.Y. and Lemmon, M.A. (1998) Specificity and promiscuity in phosphoinositide binding by pleckstrin homology domains. *Journal of Biological Chemistry*, **273**, 30497–30508.

40 Brock, C., Schaefer, M., Reusch, H.P., Czupalla, C., Michalke, M., Spicher, K., Schultz, G. and Nürnberg, B. (2003) Roles of Gβγ in membrane recruitment and activation of p110γ/p101 phosphoinositide 3-kinase γ. *Journal of Cell Biology*, **160**, 89–99.

41 Viard, P., Butcher, A.J., Halet, G., Davies, A., Nurnberg, B., Heblich, F. and Dolphin, A.C. (2004) PI3K promotes voltage-dependent calcium channel trafficking to the plasma membrane. *Nature Neuroscience*, **7**, 939–946.

42 Halet, G., Tunwell, R., Balla, T., Swann, K. and Carroll, J. (2002) The dynamics of plasma membrane PtdIns(4,5)P$_2$ at fertilization of mouse eggs. *Journal of Cell Science*, **115**, 2139–2149.

43 Halet, G., Tunwell, R., Parkinson, S.J. and Carroll, J. (2004) Conventional PKCs regulate the temporal pattern of Ca^{2+} oscillations at fertilization in mouse eggs. *Journal of Cell Biology*, **164**, 1033–1044.

44 Balla, T., Bondeva, T. and Várnai, P. (2000) How accurately can we image inositol lipids in living cells? *Trends in Pharmacological Sciences*, **21**, 238–241.

45 Várnai, P., Bondeva, T., Tamás, P., Tóth, B., Buday, L., Hunyady, L. and Balla, T. (2005) Selective cellular effects of overexpressed pleckstrin-homology domains that recognize PtdIns(3,4,5)P$_3$ suggest their interaction with protein binding partners. *Journal of Cell Science*, **118**, 4879–4888.

46 Hirose, K., Kadowaki, S., Tanabe, M., Takeshima, H. and Iino, M. (1999) Spatiotemporal dynamics of inositol 1,4,5-trisphosphate that underlies complex Ca^{2+} mobilization patterns. *Science*, **284**, 1527–1530.

47 Xu, C., Watras, J. and Loew, L.M. (2003) Kinetic analysis of receptor-activated phosphoinositide turnover. *Journal of Cell Biology*, **161**, 779–791.

3
The Use of Lipid-Binding Toxins to Study the Distribution and Dynamics of Sphingolipids and Cholesterol

Reiko Ishitsuka and Toshihide Kobayashi

3.1
Introduction

The biological membrane of eukaryotic cells contains more lipid species than are needed to form a simple bilayer. These lipid molecules are not homogenously distributed but, rather, serve to organize membranes into discrete, specific domains with different properties. Depending on their function, lipids are found in different particular cellular locations. Appropriate probes for lipids are excellent tools to study the functional or structural organization of lipids.

Probes for lipids are generally classified into two categories: fluorescent lipid analogs and lipid-binding molecules. Fluorescent lipid analogs have proven to be particularly useful in membrane trafficking studies. Lipid-binding molecules include small molecules such as filipin, which binds cholesterol, and proteins. Lipid-binding proteins are categorized into antibodies and toxins. Most lipid-binding toxins recognize cell surface lipids of target cells [1]. Among them, several toxins recognize with high affinity specific lipid molecules, which act as cellular receptors. For example, cholera toxin, lysenin and perfringolysin O specifically bind to glycolipid GM1, sphingomyelin and cholesterol, respectively. Aerolysin binds to GPI-anchored proteins. These characteristics make them useful probes for lipids in cellular and model membranes. Although their high affinity and high specificity to lipids is an advantage as probes, there are certain problems in using toxins as probes for lipids. First, the molecular size of most proteins is much larger than that of lipid molecules. Second, some toxins are intrinsically multivalent. Controlling probe valency is complicated and such probes may induce alteration of the membrane organization. Third, native toxins cannot be used for living cells because they induce cytolysis. For these reasons, potentially useful lipid probes need to be characterized in detail and used appropriately. Furthermore, non-toxic probes need to be developed to study lipid dynamics in living cells. In some cases, lipid organization should be determined by using more than one probe (e.g., fluorescent lipid analogs or antibodies in combination with lipid-binding toxins).

Probes and Tags to Study Biomolecular Function. Lawrence W. Miller (Ed.)
Copyright © 2008 WILEY-VCH Verlag GmbH & Co. KGaA, Weinheim
ISBN: 978-3-527-31566-6

To date, several useful toxin probes have been developed and applied to observational studies of lipid organization on membranes. This chapter focuses on lipid-binding toxins used as probes to detect sphingolipids, cholesterol and GPI-anchored proteins. Cholera toxin nowadays is a routine reagent to stain GM1, and lysenin, perfringolysin O and aerolysin have been characterized as powerful tools. Here, we introduce possible applications to detect target lipid molecules using these toxins, and describe several protocols for typical experiments.

3.2
Cholera Toxin

3.2.1
Introduction

Cholera toxin (CT) is responsible for the symptoms produced by *Vibrio cholerae* infection. CT is an AB5 hexamer consisting of a single A subunit and 5 identical B subunits. X-ray structural analyses revealed that the five B subunits form a doughnut-shaped ring, and that the toxin A subunit is composed of two distinct domains, A1 and A2. The subunit A is an ADP-ribosyltransferase, which disrupts the proper signaling of G proteins. The nontoxic five B-subunits bind with high affinity to the glycolipid receptor, monosialoganglioside, GM1 (Galβ1,3GalNAcβ1,4(NeuAcα2,3)Galβ1,4GlcCer), found in the plasma membranes of mammalian cells. The B pentamer components of CT are postulated to be carrier molecules, principally involved in delivering the toxin A-subunit into cells. CT enter host cells by moving retrograde in the secretory pathway from the plasma membrane to the endoplasmic reticulum (ER) (reviewed in Refs. [2–4]).

The interaction between the cholera toxin B subunit (CTB) and GM1 has been investigated using various techniques such as solid-phase and TLC overlay assays, surface plasmon biosensing, fluorescence spectroscopy, atomic force microscopy (AFM), and isothermal titration calorimetry. These studies have revealed that CTB binds to GM1 with remarkably high affinity (Kd $<10^{-9}$ M *in vitro*). In addition, CTB has been found to bind, with lower affinity, to a number of other gangliosides. These include, in order of binding affinity: GD1b, asialoGM1, GQ1b, GD1a, GT1b, and GM2. Studies in ganglioside-deficient cell lines have provided evidence that GM1 is the functional receptor for toxic activity. The X-ray structure of the CTB pentamer complexed with GM1 pentasaccharide revealed that each B subunit has a GM1 binding pocket, with the B subunits interacting mainly with the terminal galactose and, to a lesser extent, with the sialic acid and N-acetylgalactosamine of GM1. This study indicates that the stoichiometry of CTB: GM1 is 1 : 1, that is, a pentamer of CTB interacts with five molecules of GM1 (reviewed in Ref. [3]).

Since CTB specifically and potently binds GM1, CT as well as CTB are routinely used to stain GM1 on the cellular membranes. In addition, due to its non-toxicity, CTB can be used in living cells. To date, CTB and CT have been mainly used for studies on (i) CT function and (ii) the distribution and dynamics of GM1 or GM1-

containing microdomains. GM1 is often used as a marker of "lipid rafts", lipid domains classically considered as enriched with sphingolipids and cholesterol. Lipid rafts are proposed to play important roles in cell-signaling events and intracellular trafficking. The function of CT and the distribution of GM1 are always closely interrelated because the toxicity of CT is dependent on GM1-binding activity. Here, we focus on the use of CTB to study (i) lipid rafts (GM1-rich microdomains) on the cell surface and (ii) the intracellular trafficking of CT.

3.2.2
CTB as a Tool to Study Cell Surface Lipid Rafts

The addition of cholesterol to model membranes containing sphingolipids and glycerolipids with unsaturated fatty acid, induces the phase separation of two immiscible liquid-ordered (lipid rafts) and liquid-disordered phases. GM1 is preferentially partitioned into liquid-ordered domains, as visualized by the binding of fluorescent CTB. Liquid-ordered domains (lipid rafts) are recovered in detergent-resistant membrane (DRM) fraction [5, 6]. The association of CT with DRM has been reported many times in the literature. Due to its preference for liquid-ordered domains and DRM, GM1 is widely accepted as a good marker for lipid rafts.

3.2.2.1 Use of CTB for Biophysical Characterization of Lipid Rafts

To assess the presence as well as the heterogeneity of lipid rafts in the membranes, various biophysical techniques have been employed using CTB as a probe. These techniques include spectroscopic measurements such as fluorescence resonance energy transfer (FRET) and diffusion measurement techniques including fluorescence recovery after photobleaching (FRAP), fluorescence correlation spectroscopy (FCS), and single-particle tracking (SPT). Due to the complexity and dynamics of these membrane microdomains, controversial results have been published concerning the presence and size of lipid rafts. FRET measurement of different GPI-anchored proteins and GM1 labeled with CTB on cellular membranes detected no clustering at all, suggesting that lipid rafts were vanishingly small or that the density of the marker molecule is too low to support intermolecular FRET [7, 8]. In contrast, another FRET study on CTB-labeled GM1 and GPI-anchored proteins on the cell surface concluded that the molecules are clustered at the submicron level [9]. Using FRAP, Kenworthy *et al.* measured the diffusion of CTB-labeled GM1 and fluorescent GPI-anchored proteins on the membranes. They observed the slowest diffusion of CTB, which could potentially arise from cross-linking and trapping of other cell surface proteins interacting with the cytoskeleton, in addition to the presence of membrane microdomains [10]. FCS can easily distinguish CTB bound to GM1 (a raft marker) and dialkylcarbocyanine dye diI (a nonraft marker) by their different diffusional mobilities in both cells and domain-exhibiting model membranes. CTB was found to be immobile in cell membranes, disruption of the cytoskeleton was required to achieve higher mobility [11]. Recently, the dynamic confinement of membrane microdomain components has been proposed. Using SPT analysis, signaling lipid rafts were identified as transient confinement zones (TCZs), where a protein or lipid is confined much longer than would be expected by simple Brownian motion. SPT analysis using gold-conjugated CTB showed that raft markers including GM1 and GPI-anchored proteins spend more time in TCZs than do non-raft lipid analogs [12, 13].

3.2.2.2 Electron Microscopic Studies Using CTB

Electron microscopy has been applied to the search for a much more complex topographical organization of membrane domains. Using plasma membrane sheets of resting mast cells, Wilson *et al.* reported that GM1 labeled with CTB was uniformly distributed but that FcεRI and Thy-1 (GPI-anchored protein) were found predominantly in separate 20–50 nm diameter microdomains. External cross-linking of GM1 causes their redistribution in microdomains (>500 nm in diameter) [14]. Another electron microscopic study using CTB and lysenin showed that GM1 and sphingomyelin form distinct microdomains (120–140 nm diameter) on plasma membranes of Jurkat cells [15]. By immunoelectron microscopy using quick-frozen and freeze-fractured specimens, Fujita *et al.* recently demonstrated that GM1 (labeled with anti-GM1 antibodies as well as CTB) and GM3 form clusters that are susceptible to cholesterol depletion and chilling [16]. The clusters of GM1 and GM3 were separated in most cases, suggesting the presence of heterogeneous microdomains.

Thus, CTB is a useful tool to study the dynamics and organization of GM1 and GM1-rich microdomains (lipid rafts) on the cell surface. One of the drawbacks of CTB as a lipid probe is its multivalency. As described above, CTB is a pentameric molecule that can bind to five GM1 molecules. Because of this multivalency, the binding of CTB to GM1 could induce cross-linking and relocalization of GM1 and, subsequently, enhance the partitioning of GM1 (now a toxin-GM1 complex) into lipid raft microdomains. Furthermore, lipid rafts might be artificially induced or stabilized by the cross-linking of GM1. This potential artifact originating from CTB multivalency always has to be considered when employing CTB as a probe.

3.2.3
Intracellular Trafficking of CT

CTB has been proven to be useful not only to detect GM1-enriched domains at the plasma membrane, but also to follow the retrograde lipid trafficking from the plasma membrane to the Golgi and the ER. CT enters host cells by moving retrograde along the secretory pathway from the plasma membrane to the ER. After binding to GM1 on the plasma membranes of target cells, CT can enter cells by numerous modes of endocytosis. Early studies found caveolae to be important in CT endocytosis, and CT was considered to be a marker for caveolar endocytosis due to the abundance of GM1 in caveolae. However, it is now recognized that CT entry can occur via both clathrin-dependent and clathrin-independent (i.e., caveolar) mechanisms. CT can also enter cells by a noncaveolae and nonclathrin-mediated pathway. It is speculated that membrane lipid composition is a crucial determinant for the specific contribution of the different endocytic pathways. Following cell entry, the toxin is delivered to early endosomes, possibly through the GPI-anchored protein-enriched endosomal compartment. CT is then transported from early endosomes to the Golgi. This step is inhibited by brefeldin A, which disables COPI- and COPII-mediated vesicular transport. The transport from the Golgi to the ER is not dependent on a KDEL retrieval motif in the A subunit of the toxin, but is facilitated by it. Once transported to the ER, ER-resident chaperones and enzymes can facilitate reduction of the internal disulfide bonds and preparation of the A1 subunit for transport into the cytosol by Sec61p complex. On reaching the cytoplasm, the A1 subunit avoids proteasomal mediated degradation and catalyzes ADP-ribosylation of the heterotrimeric G proteins Gsα (reviewed in Refs. [2, 17, 18]).

The relationship between lipid rafts and CT trafficking has been studied (reviewed in Refs. [4, 18, 19]). For AB5 toxins such as Shiga toxin, toxin binding to the glycolipid receptors associated with DRMs correlates with toxin transport into the ER and its toxic activity. This is also true for CT. CT binds to GM1 associated with DRMs, then moves to the ER and induces toxicity. Cholesterol levels can modify the trafficking and function of CT. Cholesterol depletion results in slower internalization and attenuated toxicity of CT, but the CT-GM1 complex appears to remain associated with DRMs. This implies that cholesterol is required for association of the CT-GM1 complex with other lipid raft components important for raft dynamics and trafficking. The actin cytoskeleton can also play an important role in CT toxicity. Depolymerization or stabilization of actin filaments inhibits CT trafficking from the plasma membrane to Golgi and toxin functionality. However, the role of lipid rafts in the association between the CT-GM1 complex and the actin cytoskeleton remains unknown. Thus, it seems that CT endocytosis and induction of toxicity are dependent on the binding of the toxin to lipid raft domains containing GM1 and cholesterol.

3.2.4
Protocols

3.2.4.1 Materials
Alexa Fluor 488-labeled CTB was purchased from Invitrogen (Carlsbad, CA). Other CTB derivatives such as Alexa Fluor 546-, Alexa Fluor 555-, Alexa Fluor 594- and Alexa Fluor 647- conjugates of CTB as well as biotin and horseradish peroxidase conjugates were provided by Invitrogen. All the CTB conjugates from Invitrogen are prepared from recombinant CTB, which is completely free of the toxic subunit A, thus eliminating any concern about toxicity or ADP-ribosylating activity.

3.2.4.2 Observation of Trafficking of GM1 in Living Cells Using CTB

Steps

1. Grow Hela cells on 35 mm glass-bottom culture dishes (Iwaki, Chiba, Japan) for 2 d to reach 50–80% confluency.
2. Wash cells twice with ice-cold DMEM-F12.
3. Incubate cells for 15 min on ice with DMEM-F12 containing $10 \, \mu g \, ml^{-1}$ Alexa Fluor 488-labeled CTB.

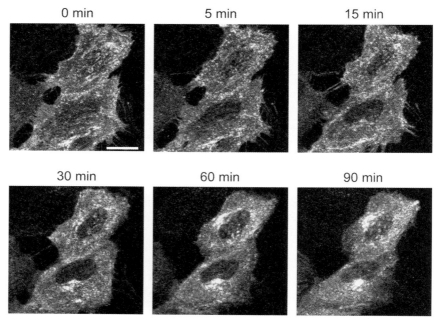

Figure 3.1 Trafficking of GM1 in Hela cells. Hela cells were incubated with Alexa Fluor 488-labeled CTB on ice for 15 min and further incubated at 37 °C for various times. Bar, 20 μm.

4. Wash cells twice with ice-cold DMEM-F12.
5. Add prewarmed DMEM-F12 and transfer cells to the microscope stage prewarmed at 37 °C.
6. Acquire images at appropriate intervals under a confocal microscope at 37 °C (Figure 3.1).

Comments

- Alexa Fluor 488-labeled CTB first binds to GM1 on the plasma membrane (Figure 3.1, 0 min). CT-GM1 complex is then rapidly internalized into the endosomal compartment (5–10 min). CT is then transported to the Golgi (30–90 min).

- To avoid rapid internalization of CTB, cells are labeled with the probe at 4 °C.

- Fluorescent-labeled GM1 has also been used to follow GM1 trafficking. C5-BODIPY GM1 has been shown to be excluded from lipid raft domains in model membranes [20, 21], probably because of the bulkiness of the BODIPY group. This is in contrast to the partition of native GM1 visualized by CTB [21]. However, BODIPY-GM1 is internalized in a clathrin-independent, caveolin-1 mediated endocytic pathway similar to that taken by cholera toxin [22].

- When using CTB as a probe, one has to take into consideration the effects of the multivalency of CTB (see above).

3.3
Lysenin

3.3.1
Introduction

Lysenin is a pore-forming toxin derived from the coelomic fluid of the earthworm *E. foetida* [23]. Lipid-binding analysis using ELISA and TLC methods indicate that this toxin binds specifically to sphingomyelin, but not to other lipids [24]. Sphingomyelin-containing liposomes specifically inhibit lysenin-induced hemolysis [24]. Upon binding to sphingomyelin-containing membranes, lysenin assembles into oligomers with the subsequent formation of pores with a diameter of approximately 3 nm in target membranes. [25]. In the presence of sphingomyelin, the tryptophan fluorescence of lysenin increases and the wavelength of maximum emission undergoes a blue shift [25], indicating migration of the lysenin tryptophan residues to a less polarized environment. The crystal structure of lysenin has not been resolved and the precise processes involved in the assembly and folding of lysenin remain unclear. In addition to lysenin, several pore-forming toxins have been reported to interact with sphingomyelin. They include equinatoxin II from the sea anemone *Actinia equine* [26], Sticholysin I and II from *Stichodactyla helianthus* [27], *Vibrio cholerae* cytolysin [28], and Eiseniapore from the earthworm *E. foetida* [29]. However, the binding of these toxins to sphingomyelin is not as specific as that of lysenin. Recently, a sphingomyelin-specific cytolysin, pleurotolysin, was purified from the basidiocarps of *Pleurotus ostreatus* and has been studied [30, 31].

Using lysenin as a sphingomyelin-specific probe, the steady-state distribution of sphingomyelin can be investigated. In Figure 3.2, normal and Niemann-Pick type A (NPA) fibroblasts were fixed, permeabilized and labeled with maltose binding protein conjugated lysenin (MBP-lysenin) (See protocol 3.3.4.2). NPA cells are characterized by a deficiency in lysosomal acid sphingomyelinase and hence display intracellular accumulation of sphingomyelin [32]. Whereas in MBP-lysenin labeled small intracellular vesicles in normal cells, NPA cells displayed bright perinuclear MBP-lysenin labeling.

3.3.2
Lysenin Binds Clustered Sphingomyelin

A recent study indicates that, in addition to its sphingomyelin specificity, the binding of lysenin to sphingomyelin is dependent on the local density of the lipid, that is, lysenin binds sphingomyelin only when the lipid forms clusters [33]. In model membrane studies, lysenin binds to liposomes composed of sphingomyelin and liquid-crystalline lipids such as dioleoylphosphatidylcholine (diC18:1PC). In these membranes, the two lipids are phase-separated and sphingomyelin forms clusters. In contrast, lysenin does not bind to liposomes containing sphingomyelin and ordered lipids such as dipalmitoylphosphatidylcholine (diC16:0 PC) or glycosphingolipids. Lipids are well mixed in these type of membranes and thus the local

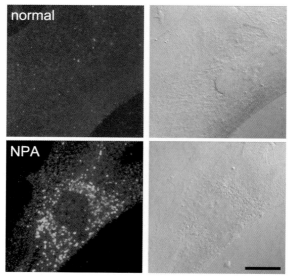

Figure 3.2 Cellular distribution of sphingomyelin labeled with MBP-lysenin. Normal and Niemann-Pick type A human skin fibroblasts were fixed, permeabilized, and incubated with MBP-lysenin. The cells were further incubated with anti-MBP antibody and Alexa Fluor 488-conjugated anti-rabbit antibody. Bar, 20 µm.

density of sphingomyelin is decreased. The fact that most mammalian cells are sensitive to lysenin suggests that sphingomyelin forms clusters in most biomembranes. There are a few exceptions, such as the apical membranes of epithelial cells and the plasma membrane of melanoma cells. Although these membranes contain sphingomyelin, they are resistant to lysenin. These membranes are highly enriched with glycosphingolipids, which can mix well with sphingomyelin and decrease its local density. Therefore, it is noteworthy that the lack of lysenin binding does not necessarily indicate a deficiency of sphingomyelin on cellular membranes.

3.3.3
Non-Toxic Lysenin as a Sphingomyelin Probe

Due to its toxicity, full-length lysenin cannot be used as a sphingomyelin probe in living cells. To follow the distribution and dynamics of sphingomyelin in intact cells, a non-toxic form of lysenin is required. Studies using truncated lysenin mutants have revealed that the N-terminus of the protein is necessary for cytotoxicity, whereas the C-terminus is required to bind sphingomyelin [15]. The 137 amino acids of the C-terminus of the protein are sufficient to bind sphingomyelin. The minimal C-terminal fragment (amino acids 161–297) can be utilized as a non-toxic, sphingomyelin probe. This "non-toxic lysenin" (NT-Lys) does not oligomerize after binding to sphingomyelin-containing membranes. NT-Lys displays the same selectivity for sphingomyelin-containing membranes *in vitro* as full-length lysenin [15]. NT-Lys and native lysenin exhibit a comparable on-rate of binding to sphingomyelin when measured by surface plasmon resonance. In contrast, dissociation of NT-Lys is 100 times faster than that of full-length lysenin. The addition of various tags (GFP-derived fluorescent protein Venus, mRFP, glutathione-S-transferase (GST), and MBP) at the N-terminal of NT-Lys does not alter the binding specificity. Using NT-Lys, sphingomyelin on the cell surface of living cells can be labeled (See protocol 3.3.4.3). Plasma membranes of living Jurkat cells were uniformly labeled with monomeric Venus NT-Lys (Figure 3.3A). In Figure 3.3B, living Jurkat cells were doubly labeled with CTB and NT-Lys followed by fixation, the distribution of the toxins on the plasma membrane sheet was then examined by immunoelectron microscopy. Both sphingomyelin and GM1 form domains with a radius of 60–70 nm in Jurkat cells. However, there was no co-cluster of sphingomyelin and GM1 in the plasma membrane of the cells. Thus, the sphingomyelin-rich and GM1-rich domains are spatially distinct. Furthermore, this spatial heterogeneity could be related to specific sphingomyelin-dependent signaling pathways distinct from the pathways linked to TCR activation or GM1 clustering. Like T cell receptor activation and cross-linking of GM1, the cross-linking of sphingomyelin induced calcium influx and ERK phosphorylation in the cell. However, unlike CD3 or GM1 clustering, the cross-linking of sphingomyelin did not induce significant protein

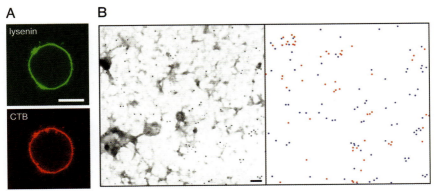

Figure 3.3 Cell surface distribution of sphingomyelin-rich domains and GM1-rich domains on the plasma membrane. (A) Jurkat cells were incubated with monomeric Venus-NT-Lys in the presence of Alexa Fluor 594-conjugated CTB. Living cell images were obtained by confocal microscopy. Bar, 10 μm; (B) Jurkat cells were first labeled with monomeric Venus-NT-Lys and biotinylated CTB at 4 °C and then fixed with 4% paraformaldehyde and 0.02% glutaraldehyde for 10 min at 4 °C. The fixed cells were labeled with anti-GFP rabbit polyclonal antibody at 4 °C followed by labeling with goat anti-rabbit IgG-5-nm gold and goat anti-biotin IgG-10-nm gold at 4 °C. The distribution of gold particles on the plasma membrane was examined under electron microscopy after ripping off the membrane. Bar, 100 nm. In the right panel, distribution of sphingomyelin (5 nm gold) is colored in red, whereas the distribution of GM1 (10 nm gold) is in blue (from Ref. [15]).

tyrosine phosphorylation. Sphingomyelinase treatment of Jurkat cells abolished LPA-mediated but not TCR-dependent signal transduction. These results suggest that the sphingomyelin-rich domain provides a functional signal cascade platform that is distinct from those provided by the T cell receptor or GM1 [15].

3.3.4
Protocols

3.3.4.1 Materials

Lysenin and anti-lysenin antibody were purchased from Peptide Institute Inc. (Osaka, Japan). MBP-lysenin was expressed in *E. coli* and purified using a maltose column according to the manufacturer's protocol (New England Biolabs, Beverly, MA). His-tagged Venus NT-Lys and His-tagged monomeric red fluorescent protein (mRFP) NT-Lys were expressed in *E. coli* and lysed and purified using nickel columns according to the manufacturer's protocol (QIAGEN). We used monomer Venus (replacing amino acid Ala-206 with Lys) to avoid dimerization of the Venus protein. Anti-MBP antibody was from New England BioLabs. Anti-rabbit antibody conjugated to Alexa Fluor 488 was purchased from Invitrogen (Carlsbad, CA). Mowiol was purchased from Aldrich chemical (Milwaukee, WI).

3.3.4.2 Cellular Staining of Sphingomyelin by Lysenin

Steps

All manipulations are performed at room temperature unless otherwise noted.

1. Grow cells on coverslips for 2 d to reach 50–80% confluency.
2. Fix cells with 3% paraformaldehyde in phosphate-buffered saline (PBS) for 20 min.
3. Incubate cells with 50 mM NH_4Cl in PBS for 10 min.
4. Permeabilize cells with 50 µg ml^{-1} digitonin for 10 min.
5. Incubate cells with 0.2% gelatin in PBS for 20 min.
6. Incubate cells with 1 µg ml^{-1} lysenin (or 5 µg ml^{-1} MBP-lysenin) for 1 h on ice and fix cells again.
7. Incubate cells with anti-lysenin antibody (or anti-MBP antibody) for 30 min.
8. Incubate cells with anti-rabbit antibody conjugated to Alexa Fluor 488 for 30 min Wash cells with PBS
9. Mount specimens in Mowiol.
10. Acquire images under confocal microscopy (Figure 3.2).

Comments

- To confirm the specific binding of lysenin to sphingomyelin, we recommend examining whether sphingomyelinase treatment of the cells abolishes the lysenin-labeling. To perform sphingomyelinase treatment, cells are incubated with 10 mU ml^{-1} recombinant *B. cereus* sphingomyelinase (Sigma) in PBS for 1 h at 37 °C after permeabilization of the cells (between steps 4 and 5).

- To avoid oligomerization of lysenin, we recommend incubating cells with lysenin on ice (or at 4 °C) and fixing of the cells once again.

3.3.4.3 Observation of Sphingomyelin on the Plasma Membrane of Living Cells Using Non-Toxic Lysenin

Steps

1. Wash Jurkat cells twice with ice-cold RPMI 1640.
2. Incubate cells for 20 min on ice with RPMI 1640 containing $20\,\mu g\,ml^{-1}$ monomeric Venus-NT-Lys.
3. Wash cells with ice-cold RPMI-1640.
4. Place cells onto a glass-bottomed culture dish precoated with poly-L-lysine.
5. Acquire images under confocal microscopy (Figure 3.3A).

Comments

- To avoid rapid internalization of NT-Lys, cells are labeled with the probe on ice (or at 4 °C).

3.4 Perfringolysin O

3.4.1 Introduction

Perfringolysin O (PFO, also known as θ toxin) is a cholesterol-binding, pore-forming toxin produced by *Clostridium perfringens*. It belongs to a large family of cholesterol-dependent cytolysins (CDCs, also known as "thiol-activated toxins" or "cholesterol-binding toxins") which contribute to the pathogenesis of a large number of Gram-positive bacterial pathogens. There are more than 20 members of the CDC family identified to date. For example, PFO, streptolysin O, listeriolysin and pneumolysin belong to this family. They share a high degree of homology in their amino acid sequences (40–70%), suggesting that they all have similar activities and 3D structures [34].

Among the CDC family members, PFO has been well characterized both structurally and functionally. Its cytolytic mechanism is as follows: binding to cholesterol in the membranes, self-assembly on the membrane to form oligomers consisting of approximately 50 monomers [35] and the formation of pores leading to cell lysis. The crystal structure revealed that PFO elongated rod-shaped molecules are rich in beta sheets and composed of four domains [36]. In the first step, the toxin binds to cell surface cholesterol via domain 4 [37]. Domain 4 is the smallest functional unit displaying the same cholesterol-binding activity as the full-size toxin together with structural stability [38]. Domain 1 is suggested to participate in the monomer–monomer interactions leading to oligomer formation [36]. Two amino acid stretches in domain 3 have been reported to insert into membranes to form pores [39, 40].

3.4.2
Non-Toxic Derivatives of PFO Bind Cholesterol-Rich Domains

PFO binds with great specificity and high affinity ($K_d \sim 10^{-9}$) to cholesterol in membranes. Non-toxic derivatives of PFO have been developed as a probe for membrane cholesterol. In this derivative, PFO is nicked between the 144th and 145th amino acids by limited proteolysis with Subtilisin Carsberg protease to produce a complex of two fragments. This derivative (Cθ) lacks the capacity for oligomerization and hemolytic activity below 20 °C even though it binds to cholesterol [41, 42]. Cθ [43] is either methylated or biotinylated [44] at the ε-amino groups of lysine residues and the alpha-C terminus. These products, methylated Cθ (MCθ) and biotinylated Cθ (BCθ) have the same binding specificity and the same affinity for membrane cholesterol as intact PFO, but do not inflict significant damage to membranes, even at 37 °C [43, 44]. In model membrane studies, the PFO derivatives bind only to cholesterol-enriched liposomes, whereas the binding is not observed when the amount of cholesterol is below 25 mol% [45]. The results indicate that the binding of PFO to cholesterol is dependent on the cholesterol content in the membranes.

When cells are incubated with BCθ, treated with ice-cold Triton X-100, and fractionated on sucrose-density gradients, the BCθ bound to the cells is predominantly recovered in the DRM fraction together with other raft marker molecules. Thus, BCθ prefers one of the populations of cholesterol fractionated into the DRM fraction [46]. However, an electron microscopic study indicated that not all vesicles in the DRM fraction associated with BCθ, suggesting that the DRM fraction contains heterogeneous populations with different cholesterol contents [46].

3.4.3
Use of BCθ to Detect Cholesterol-Rich Domains

Among the PFO derivatives, BCθ is particularly useful for detecting cholesterol in cellular membranes because its binding can be visualized in the cells after coupling with fluorescent avidin. In contrast to the staining with filipin, a cholesterol-binding polyene antibiotic, BCθ was found to bind to specific populations of cholesterol in the membranes of A431 cells [46]. Cellular membranes of A431 cells are uniformly labeled with BCθ and filipin under optical microscopy. When cholesterol in cells was depleted by 30% with MβCD, BCθ binding completely disappeared, whereas filipin staining still remained [46]. Thus, BCθ appeared as a useful tool for cytochemical and biochemical studies on cholesterol-rich membrane domains. Cholesterol-rich domains in platelets have been visualized using BCθ with confocal and immunoelectron microscopy. In resting platelets, the cholesterol-rich domains are uniformly distributed on the cell surface. Upon platelet activation, cholesterol-rich domains are accumulated at the extended tips of the filopodia and at the leading edge of spreading cells [47]. Immunoelectron microscopy observation of the cytosections showed strong labeling of BCθ in plasma membranes, internal vesicles of multivesicular bodies, and exosomes [48]. When the cholesterol distribution along the endocytic pathway was studied by electron microscopy, BCθ labeling was found in recycling tubulovesicles and multivesicular bodies, but not in lysosome [49].

3.4.4
Protocol

3.4.4.1 Materials

PFO (θ toxin) was expressed in *E. coli*, and purified as described [50]. PFO was digested with subtilisin Carlsberg and then biotinylated to yield BCθ [44]. Alexa Fluor 488-conjugated streptavidin was purchased from Invitrogen (Carlsbad, CA). Mowiol was purchased from Aldrich chemicals (Milwaukee, WI).

3.4.4.2 Staining of Cholesterol-Rich Domain in the Plasma Membrane Using BCθ

Steps

1. Grow cells on coverslips. Cells are incubated in a serum-free medium for 2 h to eliminate the effects of serum cholesterol.
2. Wash cells with a serum-free medium.
3. Incubate cells for 30 min on ice with $10\,\mu g\,ml^{-1}$ BCθ in binding buffer (0.1% bovine serum albumin (BSA)/10 mM HEPES pH 7.5/medium).
4. Wash cells with serum-free medium.
5. Fix cells with 3% paraformaldehyde for 20 min.
6. Incubate cells with 50 mM NH_4Cl in PBS for 10 min.
7. Incubate cells with Alexa Fluor 488-conjugated streptavidin for 30 min.
8. Wash cells with PBS.
9. Mount specimens with Mowiol.
10. Acquire images under confocal microscopy.

Comments

- To reduce background staining, cells should be preincubated with a serum-free medium.
- To confirm specific binding of BCθ to cholesterol, we recommend investigating whether MβCD treatment of the cells abolishes the staining. For MβCD treatment, cells are incubated with 5 mM MβCD in serum-free medium for 15 min at 37 °C.
- In this protocol, living cells are first labeled with BCθ on ice and then fixed with paraformaldehyde. BCθ labeling is performed on the fixed cells. When cells are first fixed, the incubation of the cells with BCθ is performed at room temperature, as previously reported [46, 51].
- BCθ can also be used to detect intracellular cholesterol-rich domains. Sugii *et al.* showed that increasing the concentration of paraformaldehyde during the fixation step induces the permeabilization of the cells without using a specific

permeabilization reagent and thus, the BCθ entry to stain the cholesterol-rich domains inside mammalian cells [51].

3.5 Aerolysin

3.5.1 Introduction

Aerolysin is one of the major toxins secreted by the Gram-negative bacterium *Aeromonas hydrophia*. Aerolysin is secreted as a soluble precursor, proaerolysin, with a typical signal sequence. Once outside the cell, proaerolysin is activated by proteolytic removal of the C-terminus. The toxin then binds to cell surface receptors of the target cells promoting the formation of amphipathic heptamers. The high affinity receptors for aerolysin are glycosylphosphatidyl (GPI)-anchored proteins [52–54]. Since aerolysin binds to GPI-anchored proteins irrespective of the peptide sequence, it is suggested that the glycan core of GPI-anchored proteins acts as a receptor for the toxin. However, since there is no binding to protein-free GPIs, the polypeptide moiety is required for the toxin–receptor interaction [55]. These oligomers then insert into the membrane to form voltage-gated channels that lead to osmotic lysis of the cells. The crystal structure revealed that aerolysin is divided into four domains, composed mainly of β-sheets [56]. Domain 1 and domain 2 are important for high affinity binding to GPI-anchored proteins [57]. Domain 4 is revealed to be in the transmembrane region and domain 3 is shown to constitute the mouth of the channel [56, 58]. A loop in domain 3, which forms a transmembrane β-hairpin, has been shown to be crucial for proper membrane insertion and is directly involved in channel formation [59].

3.5.2 Use of Aerolysin to Detect GPI-Anchored Proteins

Since aerolysin binds GPI-anchored proteins with high affinity both on cells and *in vitro*, the toxin has been used as a specific probe to study GPI-anchored proteins. Aerolysin overlay assays have been performed to detect GPI-anchored proteins in cell extracts. To observe the movement of GPI-anchored proteins in living cells, a double mutant of aerolysin has been used [53]. In this mutant, residues 202 and 445 are changed to cysteines, creating a disulfide bridge between the C-terminal propeptide and the mature toxin [53]. This mutant binds specifically to GPI-anchored proteins like the wild-type toxin, but exhibits no toxicity, even after C-terminal proteolysis, unless it becomes reduced [53]. This non-toxic aerolysin mutant conjugated with Alexa was used to monitor endocytosed GPI-anchored proteins in living cells [60]. The study showed that internalized GPI-anchored proteins are delivered to early endosome antigen (EEA) 1-positive endosomes, *en route* to late endosomes in the baby hamster kidney (BHK) cells [60].

In addition to aerolysin, fluorescent antibodies or ligands have been used to visualize endocytosis of specific GPI-anchored proteins. For example, fluorescent derivatives of folic acid are useful for visualization of the endocytosis of GPI-anchored folate receptor [61, 62]. GFP-tagged GPI-anchored proteins or GFP-GPI have been widely used to monitor GPI-anchored proteins in living cells. The organization of GPI-anchored proteins on biological membranes has been analyzed by biophysical studies such as FRET [63] and FRAP [10] using GFP-labeled GPI-anchored proteins. To observe the endocytosis of GFP-tagged GPI (-anchored proteins), cycloheximide should be used to inhibit protein synthesis and to remove any contribution from biosynthetic trafficking in the cells.

References

1 Reig, N. and van der Goot, F.G. (2006) About lipids and toxins. *FEBS Letters*, **580**, 5572–5579.

2 Lencer, W.I. and Tsai, B. (2003) The intracellular voyage of cholera toxin: going retro. *Trends in Biochemical Sciences*, **28**, 639–645.

3 De Haan, L. and Hirst, T.R. (2004) Cholera toxin: a paradigm for multi-functional engagement of cellular mechanisms (Review). *Molecular Membrane Biology*, **21**, 77–92.

4 Lencer, W.I. and Saslowsky, D. (2005) Raft trafficking of AB5 subunit bacterial toxins. *Biochimica et Biophysica Acta*, **1746**, 314–321.

5 Edidin, M. (2003) The state of lipid rafts: from model membranes to cells. *Annual Review of Biophysics and Biomolecular Structure*, **32**, 257–283.

6 Simons, K. and Vaz, W.L. (2004) Model systems, lipid rafts, and cell membranes. *Annual Review of Biophysics and Biomolecular Structure*, **33**, 269–295.

7 Kenworthy, A.K. and Edidin, M. (1998) Distribution of a glycosylphosphatidylinositol-anchored protein at the apical surface of MDCK cells examined at a resolution of <100 Å using imaging fluorescence resonance energy transfer. *The Journal of Cell Biology*, **142**, 69–84.

8 Kenworthy, A.K., Petranova, N. and Edidin, M. (2000) High-resolution FRET microscopy of cholera toxin B-subunit and GPI- anchored proteins in cell plasma membranes. *Molecular Biology of the Cell*, **11**, 1645–1655.

9 Nichols, B.J. (2003) GM1-containing lipid rafts are depleted within clathrin-coated pits. *Current Biology*, **13**, 686–690.

10 Kenworthy, A.K., Nichols, B.J., Remmert, C.L., Hendrix, G.M., Kumar, M., Zimmerberg, J. and Lippincott-Schwartz, J. (2004) Dynamics of putative raft-associated proteins at the cell surface. *The Journal of Cell Biology*, **165**, 735–746.

11 Bacia, K., Scherfeld, D., Kahya, N. and Schwille, P. (2004) Fluorescence correlation spectroscopy relates rafts in model and native membranes. *Biophysical Journal*, **87**, 1034–1043.

12 Sheets, E.D., Lee, G.M., Simson, R. and Jacobson, K. (1997) Transient confinement of a glycosylphosphatidylinositol-anchored protein in the plasma membrane. *Biochemistry*, **36**, 12449–12458.

13 Dietrich, C., Yang, B., Fujiwara, T., Kusumi, A. and Jacobson, K. (2002) Relationship of lipid rafts to transient confinement zones detected by single particle tracking. *Biophysical Journal*, **82**, 274–284.

14 Wilson, B.S., Steinberg, S.L., Liederman, K., Pfeiffer, J.R., Surviladze, Z., Zhang, J., Samelson, L.E., Yang, L.H., Kotula, P.G. and Oliver, J.M. (2004) Markers for detergent-resistant lipid rafts occupy distinct and dynamic domains in native membranes. *Molecular Biology of the Cell*, **15**, 2580–2592.

15 Kiyokawa, E., Baba, T., Otsuka, N., Makino, A., Ohno, S. and Kobayashi, T. (2005) Spatial and Functional Heterogeneity of Sphingolipid-rich Membrane Domains. *Journal of Biological Chemistry*, **280**, 24072–24084.

16 Fujita, A., Cheng, J., Hirakawa, M., Furukawa, K., Kusunoki, S. and Fujimoto, T. (2007) Gangliosides GM1 and GM3 in the Living Cell Membrane Form Clusters Susceptible to Cholesterol Depletion and Chilling. *Molecular Biology of the Cell*, **18**, 2112–2122.

17 Sandvig, K. and van Deurs, B. (2002) Transport of protein toxins into cells: pathways used by ricin, cholera toxin and Shiga toxin. *FEBS Letters*, **529**, 49–53.

18 Chinnapen, D.J., Chinnapen, H., Saslowsky, D. and Lencer, W.I. (2007) Rafting with cholera toxin: endocytosis and trafficking from plasma membrane to ER. *FEMS Microbiology Letters*, **266**, 129–137.

19 Smith, D.C., Lord, J.M., Roberts, L.M. and Johannes, L. (2004) Glycosphingolipids as toxin receptors. *Seminars in Cell & Developmental Biology*, **15**, 397–408.

20 Wang, T.Y. and Silvius, J.R. (2000) Different sphingolipids show differential partitioning into sphingolipid/cholesterol-rich domains in lipid bilayers. *Biophysical Journal*, **79**, 1478–1489.

21 Shaw, J.E., Epand, R.F., Epand, R.M., Li, Z., Bittman, R. and Yip, C.M. (2006) Correlated fluorescence-atomic force microscopy of membrane domains: structure of fluorescence probes determines lipid localization. *Biophysical Journal*, **90**, 2170–2178.

22 Singh, R.D., Puri, V., Valiyaveettil, J.T., Marks, D.L., Bittman, R. and Pagano, R.E. (2003) Selective caveolin-1-dependent endocytosis of glycosphingolipids. *Molecular Biology of the Cell*, **14**, 3254–3265.

23 Sekizawa, Y., Hagiwara, K., Nakajima, T. and Kobayashi, H. (1996) A novel protein, lysenin, that causes contraction of the isolated rat aorta: its purification from the coelomic fluid of the earthworm, Eisenia foetida. *Biomedical Research*, **17**, 197–203.

24 Yamaji, A., Sekizawa, Y., Emoto, K., Sakuraba, H., Inoue, K., Kobayashi, H. and Umeda, M. (1998) Lysenin a novel sphingomyelin-specific binding protein. *Journal of Biological Chemistry*, **273**, 5300–5306.

25 Yamaji-Hasegawa, A., Makino, A., Baba, T., Senoh, Y., Kimura-Suda, H., Sato, S.B., Terada, N., Ohno, S., Kiyokawa, E., Umeda, M. and Kobayashi, T. (2003) Oligomerization and pore formation of a sphingomyelin-specific toxin, lysenin. *Journal of Biological Chemistry*, **278**, 22762–22770.

26 Macek, P., Zecchini, M., Pederzolli, C., Dalla Serra, M. and Menestrina, G. (1995) Intrinsic tryptophan fluorescence of equinatoxin II, a pore-forming polypeptide from the sea anemone Actinia equina L, monitors its interaction with lipid membranes. *European Journal of Biochemistry*, **234**, 329–335.

27 Valcarcel, C.A., Dalla Serra, M., Potrich, C., Bernhart, I., Tejuca, M., Martinez, D., Pazos, F., Lanio, M.E. and Menestrina, G. (2001) Effects of lipid composition on membrane permeabilization by sticholysin I and II, two cytolysins of the sea anemone Stichodactyla helianthus. *Biophysical Journal*, **80**, 2761–2774.

28 Zitzer, A., Zitzer, O., Bhakdi, S. and Palmer, M. (1999) Oligomerization of Vibrio cholerae cytolysin yields a pentameric pore and has a dual specificity for cholesterol and sphingolipids in the target membrane. *Journal of Biological Chemistry*, **274**, 1375–1380.

29 Lange, S., Nussler, F., Kauschke, E., Lutsch, G., Cooper, E.L. and Herrmann, A. (1997) Interaction of earthworm hemolysin with lipid membranes requires sphingolipids. *Journal of Biological Chemistry*, **272**, 20884–20892.

30 Sakurai, N., Kaneko, J., Kamio, Y. and Tomita, T. (2004) Cloning, expression, and pore-forming properties of mature and precursor forms of pleurotolysin, a

sphingomyelin-specific two-component cytolysin from the edible mushroom Pleurotus ostreatus. *Biochimica et Biophysica Acta*, **1679**, 65–73.

31 Tomita, T., Noguchi, K., Mimuro, H., Ukaji, F., Ito, K., Sugawara-Tomita, N. and Hashimoto, Y. (2004) Pleurotolysin, a novel sphingomyelin-specific two-component cytolysin from the edible mushroom Pleurotus ostreatus, assembles into a transmembrane pore complex. *Journal of Biological Chemistry*, **279**, 26975–26982.

32 Kolodny, E.H. (2000) Niemann-Pick disease. *Current Opinion in Hematology*, **7**, 48–52.

33 Ishitsuka, R., Yamaji-Hasegawa, A., Makino, A., Hirabayashi, Y. and Kobayashi, T. (2004) A lipid-specific toxin reveals heterogeneity of sphingomyelin-containing membranes. *Biophysical Journal*, **86**, 296–307.

34 Gilbert, R.J. (2002) Pore-forming toxins. *Cellular and Molecular Life Sciences*, **59**, 832–844.

35 Olofsson, A., Hebert, H. and Thelestam, M. (1993) The projection structure of perfringolysin O (Clostridium perfringens theta-toxin). *FEBS Letters*, **319**, 125–127.

36 Rossjohn, J., Feil, S.C., McKinstry, W.J., Tweten, R.K. and Parker, M.W. (1997) Structure of a cholesterol-binding, thiol-activated cytolysin and a model of its membrane form. *Cell*, **89**, 685–692.

37 Ramachandran, R., Heuck, A.P., Tweten, R.K. and Johnson, A.E. (2002) Structural insights into the membrane-anchoring mechanism of a cholesterol-dependent cytolysin. *Nature Structural Biology*, **9**, 823–827.

38 Shimada, Y., Maruya, M., Iwashita, S. and Ohno-Iwashita, Y. (2002) The C-terminal domain of perfringolysin O is an essential cholesterol-binding unit targeting to cholesterol-rich microdomains. *European Journal of Biochemistry*, **269**, 6195–6203.

39 Shatursky, O., Heuck, A.P., Shepard, L.A., Rossjohn, J., Parker, M.W., Johnson, A.E. and Tweten, R.K. (1999) The mechanism of membrane insertion for a cholesterol-dependent cytolysin: a novel paradigm for pore-forming toxins. *Cell*, **99**, 293–299.

40 Shepard, L.A., Shatursky, O., Johnson, A.E. and Tweten, R.K. (2000) The mechanism of pore assembly for a cholesterol-dependent cytolysin: formation of a large prepore complex precedes the insertion of the transmembrane beta-hairpins. *Biochemistry*, **39**, 10284–10293.

41 Iwamoto, M., Nakamura, M., Mitsui, K., Ando, S. and Ohno-Iwashita, Y. (1993) Membrane disorganization induced by perfringolysin O (theta-toxin) of Clostridium perfringens-effect of toxin binding and self-assembly on liposomes. *Biochimica et Biophysica Acta*, **1153**, 89–96.

42 Ohno-Iwashita, Y., Iwamoto, M., Mitsui, K., Ando, S. and Nagai, Y. (1988) Protease-nicked theta-toxin of Clostridium perfringens, a new membrane probe with no cytolytic effect, reveals two classes of cholesterol as toxin-binding sites on sheep erythrocytes. *European Journal of Biochemistry*, **176**, 95–101.

43 Ohno-Iwashita, Y., Iwamoto, M., Ando, S., Mitsui, K. and Iwashita, S. (1990) A modified theta-toxin produced by limited proteolysis and methylation: a probe for the functional study of membrane cholesterol. *Biochimica et Biophysica Acta*, **1023**, 441–448.

44 Iwamoto, M., Morita, I., Fukuda, M., Murota, S., Ando, S. and Ohno-Iwashita, Y. (1997) A biotinylated perfringolysin O derivative: a new probe for detection of cell surface cholesterol. *Biochimica et Biophysica Acta*, **1327**, 222–230.

45 Ohno-Iwashita, Y., Iwamoto, M., Ando, S. and Iwashita, S. (1992) Effect of lipidic factors on membrane cholesterol topology – mode of binding of theta-toxin to cholesterol in liposomes. *Biochimica et Biophysica Acta*, **1109**, 81–90.

46 Waheed, A.A., Shimada, Y., Heijnen, H.F., Nakamura, M., Inomata, M., Hayashi, M., Iwashita, S., Slot, J.W. and Ohno-Iwashita, Y. (2001) Selective binding of

perfringolysin O derivative to cholesterol-rich membrane microdomains (rafts). *Proceedings of the National Academy of Sciences of the United States of America*, **98**, 4926–4931.

47 Heijnen, H.F., Van Lier, M., Waaijenborg, S., Ohno-Iwashita, Y., Waheed, A.A., Inomata, M., Gorter, G., Mobius, W., Akkerman, J.W. and Slot, J.W. (2003) Concentration of rafts in platelet filopodia correlates with recruitment of c-Src and CD63 to these domains. *Journal of Thromb Haemostasis*, **1**, 1161–1173.

48 Mobius, W., Ohno-Iwashita, Y., van Donselaar, E.G., Oorschot, V.M., Shimada, Y., Fujimoto, T., Heijnen, H.F., Geuze, H.J. and Slot, J.W. (2002) Immunoelectron microscopic localization of cholesterol using biotinylated and non-cytolytic perfringolysin O. *Journal of Histochemistry and Cytochemistry*, **50**, 43–55.

49 Mobius, W., van Donselaar, E., Ohno-Iwashita, Y., Shimada, Y., Heijnen, H.F., Slot, J.W. and Geuze, H.J. (2003) Recycling compartments and the internal vesicles of multivesicular bodies harbor most of the cholesterol found in the endocytic pathway. *Traffic*, **4**, 222–231.

50 Shimada, Y., Nakamura, M., Naito, Y., Nomura, K. and Ohno-Iwashita, Y. (1999) C-terminal amino acid residues are required for the folding and cholesterol binding property of perfringolysin O, a pore-forming cytolysin. *Journal of Biological Chemistry*, **274**, 18536–18542.

51 Sugii, S., Reid, P.C., Ohgami, N., Shimada, Y., Maue, R.A., Ninomiya, H., Ohno-Iwashita, Y. and Chang, T.Y. (2003) Biotinylated theta-toxin derivative as a probe to examine intracellular cholesterol-rich domains in normal and Niemann-Pick type C1 cells. *Journal of Lipid Research*, **44**, 1033–1041.

52 Nelson, K.L., Raja, S.M. and Buckley, J.T. (1997) The glycosylphosphatidylinositol-anchored surface glycoprotein Thy-1 is a receptor for the channel-forming toxin aerolysin. *Journal of Biological Chemistry*, **272**, 12170–12174.

53 Abrami, L., Fivaz, M., Glauser, P.E., Parton, R.G. and van der Goot, F.G. (1998) A pore-forming toxin interacts with a GPI-anchored protein and causes vacuolation of the endoplasmic reticulum. *The Journal of Cell Biology*, **140**, 525–540.

54 Diep, D.B., Nelson, K.L., Raja, S.M., Pleshak, E.N. and Buckley, J.T. (1998) Glycosylphosphatidylinositol anchors of membrane glycoproteins are binding determinants for the channel-forming toxin aerolysin. *Journal of Biological Chemistry*, **273**, 2355–2360.

55 Abrami, L., Velluz, M.C., Hong, Y., Ohishi, K., Mehlert, A., Ferguson, M., Kinoshita, T. and Gisou van der Goot, F. (2002) The glycan core of GPI-anchored proteins modulates aerolysin binding but is not sufficient: the polypeptide moiety is required for the toxin-receptor interaction. *FEBS Letters*, **512**, 249–254.

56 Parker, M.W., Buckley, J.T., Postma, J.P., Tucker, A.D., Leonard, K., Pattus, F. and Tsernoglou, D. (1994) Structure of the Aeromonas toxin proaerolysin in its water-soluble and membrane-channel states. *Nature*, **367**, 292–295.

57 MacKenzie, C.R., Hirama, T. and Buckley, J.T. (1999) Analysis of receptor binding by the channel-forming toxin aerolysin using surface plasmon resonance. *Journal of Biological Chemistry*, **274**, 22604–22609.

58 Tsitrin, Y., Morton, C.J., el-Bez, C., Paumard, P., Velluz, M.C., Adrian, M., Dubochet, J., Parker, M.W., Lanzavecchia, S. and van der Goot, F.G. (2002) Conversion of a transmembrane to a water-soluble protein complex by a single point mutation. *Nature Structural Biology*, **9**, 729–733.

59 Iacovache, I., Paumard, P., Scheib, H., Lesieur, C., Sakai, N., Matile, S., Parker, M.W. and van der Goot, F.G. (2006) A rivet model for channel formation by aerolysin-like pore-forming toxins. *The EMBO Journal*, **25**, 457–466.

60 Fivaz, M., Vilbois, F., Thurnheer, S., Pasquali, C., Abrami, L., Bickel, P.E.,

Parton, R.G. and van der Goot, F.G. (2002) Differential sorting and fate of endocytosed GPI-anchored proteins. *The EMBO Journal*, **21**, 3989–4000.

61 Chatterjee, S., Smith, E.R., Hanada, K., Stevens, V.L. and Mayor, S. (2001) GPI anchoring leads to sphingolipid-dependent retention of endocytosed proteins in the recycling endosomal compartment. *The EMBO Journal*, **20**, 1583–1592.

62 Sabharanjak, S., Sharma, P., Parton, R.G. and Mayor, S. (2002) GPI-anchored proteins are delivered to recycling endosomes via a distinct cdc42-regulated, clathrin-independent pinocytic pathway. *Developmental Cell*, **2**, 411–423.

63 Sharma, P., Varma, R., Sarasij, R.C., Ira Gousset, K., Krishnamoorthy, G., Rao, M. and Mayor, S. (2004) Nanoscale organization of multiple GPI-anchored proteins in living cell membranes. *Cell*, **116**, 577–589.

4
"FlAsH" Protein Labeling

Stefan Jakobs, Martin Andresen, and Christian A. Wurm

4.1
Introduction

Fluorescence microscopy has developed into a tremendously powerful tool for the study of protein dynamics within cells in time and space. A milestone for live-cell imaging has been the discovery, cloning and heterologous expression of the green fluorescent protein (GFP) from the jellyfish *Aequorea victoria* (reviewed in [1]). Since then, the discovery of further fluorescent proteins in other species, as well as various mutagenesis approaches, has diversified the spectra and the properties of these probes [2, 3]. In the meantime fluorescent proteins have been utilized for an overwhelming number of applications in cell biology. Despite the numerous benefits of the fluorescent proteins, there are limitations for their use, the most dominant being their relatively large molecular weight of $\sim$28 kDa ($\sim$230 amino acids). Due to this size, fluorescent proteins may interfere with the correct localization, stability or functionality of their respective fusion partners.

In 1998 Roger Tsien's group developed an alternative approach for the specific labeling of proteins in living cells using the high-affinity interaction between a 6-residue peptide that includes four cysteines and membrane-permeable biarsenical molecules [4]. Since its invention, the biarsenical-tetracysteine system has undergone some notable improvements and has been employed in a number of studies.

In this chapter we describe briefly the principle of this labeling system, its major current applications and limitations, and provide information for its practical use. Several comprehensive articles describing the application of this system in mammalian cells are available [5, 6]. Here we focus on the use of the biarsenical-tetracysteine system in the budding yeast *Saccharomyces cerevisiae*, although we note that many practical aspects are transferable to other organisms as well.

Probes and Tags to Study Biomolecular Function. Lawrence W. Miller (Ed.)
Copyright © 2008 WILEY-VCH Verlag GmbH & Co. KGaA, Weinheim
ISBN: 978-3-527-31566-6

4.1.1
The Biarsenical-Tetracysteine System

This labeling system makes use of the high affinity of fluorescent compounds containing two arsenic atoms to four appropriately spaced cysteines (TetCys motif: Cys-Cys-Xaa-Xaa-Cys-Cys, where Xaa is not cysteine) (Figure 4.1A). The first utilized biarsenical dye was FlAsH (*Fluorescein Arsenical Hairpin Binder*), a derivative of the well-known fluorophore fluorescein [4]. For FlAsH, dissociation constants in the sub-nanomolar range (Table 4.1) for a peptide containing the TetCys motif have been reported. In solution, FlAsH is complexed with two 1,2-ethanedithiol (EDT) molecules. It is membrane permeable and in solution only weakly fluorescent, yet becomes brightly fluorescent upon binding to a target peptide. This improves the signal to noise ratio since, predominantly, the peptide-bound fluorophore is detected.

The DNA sequence coding for the TetCys motif may be fused to a target gene by standard techniques, resulting in fusion proteins with the TetCys motif at the N- or C-terminus or incorporated into the protein. Upon association of the biarsenical fluorophore to the binding sequence, the fluorescence develops immediately, whereas the proper folding of a fluorescent protein requires tens of minutes to hours.

Even more important, the TetCys motif is up to 25 times smaller than a fluorescent protein (~230 amino acids) (Figure 4.1B). Several examples where the protein function is less perturbed by a TetCys motif than by a fluorescent protein have been reported, including G protein-coupled receptors [7], cAMP-dependent protein kinase [8] and β-tubulin [9].

Despite its obvious merits, the biarsenical-tetracysteine system has its disadvantages, the most pressing being poor signal to noise ratios due to unwanted background staining and pronounced photobleaching. To alleviate these obstacles,

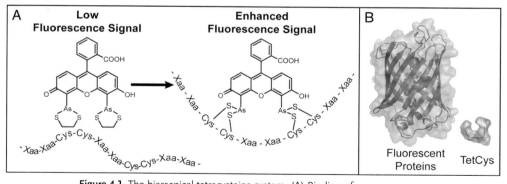

Figure 4.1 The biarsenical-tetracysteine system. (A) Binding of the biarsenical dye to the TetCys motif. Unbound FlAsH-EDT$_2$ exhibits low fluorescence, whereas bound FlAsH is brightly fluorescent. (B) Size comparison of a fluorescent protein with the TetCys motif. GFP comprises ~230 amino acid residues, whereas the TetCys motif may be as short as 6 amino acids.

Table 4.1 Binding motifs for the Biarsenical-Tetracysteine System.

Motif	Quantum efficiency	K_{on} (M^{-1} s^{-1})	K_d (pM)	Source
WEAAAREACCRECCARA	0.50	65 000	70	Griffin et al. [4]
WDCCPGCCK	0.67	310 000	4	Adams et al. [11]
WDCCGPCCK	0.44	50 000	72	Adams et al. [11]
WDCCPCCK	0.60	100 000	150	Adams et al. [11]
WDCCGCCK	0.55	35 000	100	Adams et al. [11]
WDCCDEACCK	0.23	65 000	92 000	Adams et al. [11]
FLNCCPGCCMEP	0.78	n.d.	n.d.	Martin et al. [12]
HRWCCPGCCKTF	0.65	n.d.	n.d.	Martin et al. [12]

substantial efforts have been undertaken to improve the TetCys motif, the available fluorophores and the staining procedures.

4.1.2
Improved TetCys Motifs

Utilization of the original TetCys motif (Cys-Cys-Arg-Glu-Cys-Cys) [4] generally results in comparatively strong background staining in mammalian cells. The sensitivity and the detection limit of the method were estimated to be an order of magnitude worse than those of GFP [10]. To alleviate this problem, two studies led by the Tsien lab identified several improved TetCys motifs (Table 4.1) [11, 12]. They found that the CCPGCC-FlAsH complex was five-fold more stable than the previously utilized CCRECC-FlAsH complex [11]. Subsequently, the residues flanking the core motif were also systematically optimized [12]. FLNCCPGCCMEP and HRWCCPGCCKTF were shown to tolerate washes with high concentrations of the antidote EDT while retaining the bound fluorophore. Since both sequences also resulted in higher fluorescence quantum yields, a ~20-fold increase in contrast was reported when using these improved TetCys motifs instead of the previous one. Because FLNCCPGCCMEP gives somewhat higher quantum yields than HRWCCPGCCKTF, the former is currently the TetCys motif of choice for most applications. We note, however, that for labeling applications in S. cerevisiae, we did not observe advantages of these extended TetCys motifs over CCPGCC.

4.1.3
Fluorescent Biarsenical Ligands

Although the optimization of the TetCys motif has led to important improvements in affinity and brightness, the limited photostability and pH sensitivity of FlAsH precludes many potential applications. Hence there is a large demand for improved biarsenical dyes. In 2002 a red-emitting resorufin-based analog called ReAsH (*Resorufin*-based *A*rsenical *H*airpin Binder) was introduced [11]. More recently, several other novel biarsenical ligands have been developed, some of which are listed

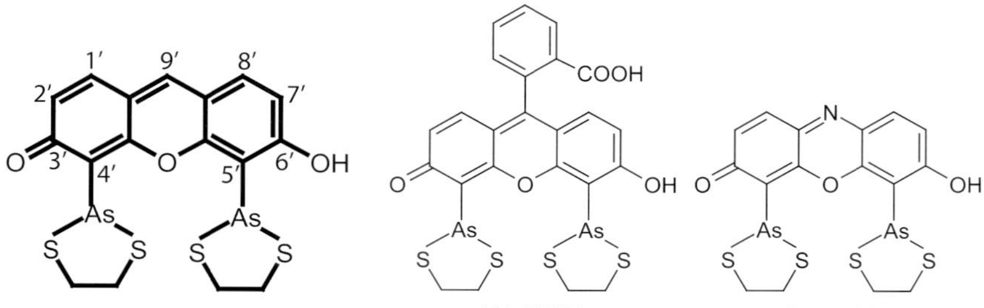

Ligand	Modification	Excitation	Emission	Source
CHoXAsH	2'-7'- Cl_2; 9'- OH	380 nm	430 nm	Adams et al. 2002
F_2FlAsH	2'-7'-F_2; 9'-(2-C_6H_4COOH)	500 nm	522 nm	Spagnuolo et al. 2006
CrAsH	9'-(2,5-C_6H_4[COOH]$_2$)	515 nm	534 nm	Cao et al. 2006
FlAsH	9'-(2-C_6H_4COOH)	511 nm	527 nm	Griffin et al. 1998
F_4FlAsH	9'-(2-C_6F_4COOH)	528 nm	544 nm	Spagnuolo et al. 2006
ReAsH	9'-Aza	593 nm	608 nm	Adams et al. 2002

Figure 4.2 Fluorescent biarsenical ligands. An overview of the chemical structures of some important TetCys motif binding probes is given. Currently only FlAsH-EDT$_2$ and ReAsH-EDT$_2$ are commercially available.

in Figure 4.2. For detailed information on the properties of these new probes, we refer the reader to [4, 11, 13, 14]. Despite promising properties, none of these novel probes have been widely used so far. Therefore it remains to be seen whether these novel ligands, or others which are still to come, will prove to be superior to FlAsH and ReAsH. Currently only these two biarsenical dyes are commercially available from Invitrogen under the brand names TC-FlAsH/Lumio green and TC-ReAsH/Lumio red.

4.1.4
Applications of the Biarsenical-Tetracysteine System

There is a remarkable variety of reported applications for this labeling system (Figure 4.3). The usability of this approach is reflected by the increasing number of studies using biarsenical dyes [7, 15–27].

Arguably, most studies exploited the small size of the TetCys motif to visualize tagged proteins within cells (Figure 4.3A). The majority of such localization applications relied on the more photostable FlAsH rather than ReAsH.

In combination the two biarsenical dyes have been used to visualize the age of TetCys tagged proteins [16, 19]. To this end, all TetCys motifs of a tagged protein in a cell are initially saturated with FlAsH. Any free dye is then washed out and the cell is allowed to synthesize new copies of the tagged protein. Subsequent staining with a biarsenical dye of different color (i.e., ReAsH) results in exclusive labeling of the newly synthesized tagged protein. Gaietta and colleagues employed this approach in an elegant experiment for the analysis of the trafficking of connexin43 [16]. They demonstrated that the newly synthesized connexin43 was predominantly incorpo-

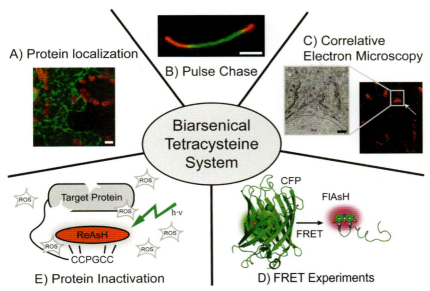

Figure 4.3 Applications of the biarsenical-tetracysteine system. Arguably, the current major application areas for the biarsenical-tetracysteine system are the following. (A) Visualization of intracellular protein localizations. Shown are FlAsH-labelled *Arabidopsis* cells expressing TetCys–labelled synthetic glycomodule peptides [22]. (B) Sequential labeling with FlAsH and ReAsH can indicate the age of proteins. This approach was used to study the life cycle of connexin43 as it was trafficked into and out of gap junctions [16]. (C) ReAsH labeled structures can be visualised in both fluorescence and electron microscopy. This correlative approach has been utilized to study Golgi vesicles throughout mitosis [24]. (D) FRET of, for example, CFP and FlAsH has been utilized to study conformational changes or to create environment-sensitive probes. (E) Intense irradiation of FlAsH or ReAsH can generate ROS, which inactivate the tagged protein, providing a genetically targeted strategy for the light driven knockout of specific proteins. All microscopic images used with permission. Scale bars (A–C): 1 μm.

rated at the periphery of existing gap junctions, whereas older connexins were removed from the center of the plaques (Figure 4.3B).

In a different application scheme, ReAsH has been demonstrated to be useful for correlative protein localization in light and electron microscopy. Upon illumination, ReAsH generates singlet oxygen species which locally oxidize diaminobenzidine (DAB) into an osmiophilic polymer which can be identified by electron microscopy. Such a photoconversion protocol has been utilized to study ReAsH labeled Golgi vesicles throughout mitosis (Figure 4.3C) [24] as well as connexin trafficking [16].

Fluorescence resonance energy transfer (FRET) between a fluorescent protein and a nearby biarsenical dye has been exploited in a few studies (Figure 4.3D). For example, the energy transfer from CFP to FlAsH was used to monitor G protein-coupled receptor activation in living cells [7]. In this study the CFP/TetCys construct showed a 3–5 times higher FRET signal than the corresponding CFP/YFP construct. In a different FRET application, a combinatorial tag consisting of GFP and a TetCys motif proved to be beneficial for photoconversion [24]. In this case, only the ReAsH bound to the TetCys motif accepts the energy from the adjacent GFP and produces singlet oxygen, thereby increasing the specificity of photoconversion.

A further field of application of the biarsenical-tetracysteine system is fluorophore assisted light inactivation of genetically targeted proteins (Figure 4.3E) [17, 18]. Here, the reactive oxygen species (ROS) that are generated by the intense illumination of FlAsH or ReAsH inactivate the tagged protein. Because of their high reactivity, the ROS act in close vicinity to the site of their generation. Thus this technique allows inactivation of tagged proteins in living cells with high spatial and temporal precision. So far only a few reports utilizing this promising technique have been published.

4.1.5
Staining in Various Model Organisms

Since its first application in cultured mammalian cells [4], the biarsenical-tetracysteine system has been successfully adapted for use in several other organisms including plants [22], budding yeast [9] and bacteria [28].

In principle most, if not all, organisms should be amenable to this labeling system, albeit various challenges may need to be faced. Physical barriers like cell walls may impede the staining procedure. Further, the system works only in a reducing environment, although this problem may be overcome by adding the membrane-permeant reducing agent tributylphosphine to the cells [24]. Also, different cell types are likely to exhibit distinct abilities to export the biarsenical dyes from the cytoplasm which will require an adaptation of the specific labeling procedure.

A major challenge in the application of the biarsenical-tetracysteine system in most cell types is nonspecific staining. One part of the unwanted background labeling appears to result from a low-affinity interaction of the biarsenical dyes with various cellular components and is frequently resolved by extensive washing. Another source of background staining, which is more difficult to address, is due to endogenous proteins containing biarsenical dye binding motifs. Although TetCys motifs of the CCXXCC type bind FlAsH with the highest affinity, motifs of the CCXCC type also display considerable binding affinities (Table 4.1). Hence endogenous proteins exhibiting CCXCC or CCXXCC sequences may represent binding partners for the biarsenical compounds. To estimate the frequency of the natural occurrence of these sequences, we performed data base searches on the sequenced genomes of some important model organisms (Table 4.2). We find that budding yeast, which displays only little unspecific background upon staining, contains relatively few endogenous potential binding partners. In this organism, there is one predicted protein which contains the CCPGCC motif; however this protein is apparently not, or only weakly expressed and thus does not contribute to an unwanted background signal. Human cells, in contrast, exhibit a higher background and also the number of potential binding partners is 10 times larger. Hence a substantial amount of the observed background staining in mammalian cells might indeed be due to endogenous proteins with a TetCys motif.

Table 4.2 Number of possible binding motifs for biarsenical dyes.

Organism	CCXCC	CCXXCC	CCPGCC
Arabidopsis thaliana	41	52	—
Candida albicans	3	3	—
Caenorhabditis elegans	38	30	—
Drosophila melanogaster	122	149	—
Homo sapiens	64	68	—
Mus musculus	58	64	—
Rattus norvegicus	27	22	—
Saccharomyces cerevisiae	5	7	1)
Saccharomyces pombe	1	1	—

4.1.6
FlAsH Labeling in *S. cerevisiae*

Functional analysis of the budding yeast *S. cerevisiae* has been greatly aided by the ease with which genes can be modified within the genome by site-specific homologous recombination. This so-called epitope-tagging allows the modification of practically any ORF without perturbing its transcription [29].

In order to utilize this method for the tagging of yeast proteins with the TetCys motif, we have previously constructed a small series of tagging vectors containing between one and three TetCys motifs (Figure 4.4) [9]. In that study the effects of the different tag lengths on the functionality of a host protein (β-tubulin; Tub2) were investigated. Tub2 tagged with either 1× TetCys (10 amino acids [aa]) or 2× TetCys (20 aa) was able to substitute Tub2. In contrast, C-terminal tagging of Tub2 with

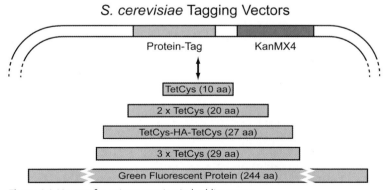

Figure 4.4 Vectors for epitope tagging in budding yeast. Schematic representation of the tagging modules (for details see [9]). The miscellaneous tagging plasmids are identical except for the sequences of the epitopes. Therefore, the same set of primers can be used to tag an ORF with any of these epitopes.

Table 4.3 Comparison of photostability.

Protein-tag	Bleaching rate, $t_{1/2}$ (s)
1× TetCys	1.8
2× TetCys	3.5
3× TetCys	9.5
TetCys-HA-TetCys	4.5
GFP	16.0

3× TetCys (29 aa) or with GFP (244 aa) resulted in nonviable haploid cells [9]. Hence, for β-tubulin, the size of the tag is crucial. Similar effects have also been reported for other proteins [7, 8], although it can be anticipated that for many host-proteins the size of the tag is only of minor relevance [30].

We compared the influence of concatenated TetCys motifs on the bleaching rate of the fluorescence signal in FlAsH labeled cells (Table 4.3). To this end we tagged the alpha subunit of the mitochondrial ATP synthase (Atp1) with the various motifs. The photobleaching rates were measured with an epi-fluorescence microscope equipped with a mercury lamp (100 W) and a CCD camera. We found the resistance against photobleaching to increase with the number of TetCys motifs attached to the host protein (Table 4.3). FlAsH labeled proteins tagged with one TetCys motif were so prone to photobleaching that we do not recommend the 1× TetCys motif for use in budding yeast. Tagging with the 2× TetCys or 3× TetCys motifs significantly reduced the bleaching rates. This reduced bleaching is likely due to FRET between the two or three FlAsH molecules binding to these motifs. Separation of two TetCys motifs by the hemagglutinin (HA) epitope (11 aa) (Figure 4.4) increases the overall brightness, as well as the resistance against photobleaching, slightly compared to the 2× TetCys motif, probably because the binding sites are better accessible for the biarsenical dyes. However, since the TetCys-HA-TetCys tag is of similar size to the 3× TetCys tag, but the latter gives rise to brighter labeled specimens (not shown), we do not recommend the TetCys-HA-TetCys tag, either. Instead, we recommend either the 2× TetCys or the 3× TetCys for most applications in budding yeast. Whenever the larger 3× TetCys motif is tolerable, it is likely to outperform the 2× TetCys motif, albeit at the expense of a slightly larger size.

Two approaches for the FlAsH-labeling of yeast cells expressing TetCys tagged proteins are currently routinely used in our laboratory (detailed protocols are given in the last section of this chapter). The first method relies on the overnight-cultivation of a liquid yeast culture in a growth medium containing FlAsH-EDT$_2$. Since FlAsH-EDT$_2$ is relatively expensive, it is advisable to use very small culture volumes which require careful adjustments of the growth conditions. Once established, this labeling protocol generally results in a homogenous labeling of the whole culture. The second approach requires a brief electric discharge provided by a standard electroporator to bring the biarsenical dyes into the cells. This approach is quick and reliable, however, generally only a small fraction (2–10%) of the cells will be labeled.

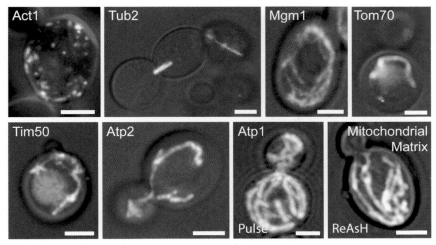

Figure 4.5 Visualization of various 2× TetCys tagged proteins in *S. cerevisiae*. Examples include the actin (Act1) and tubulin (Tub2) cytoskeleton, and different compartments of mitochondria. The tagged alpha subunit of the mitochondrial F_1F_0 ATP synthase (Atp1) was labeled using the pulse staining protocol. The mitochondrial matrix was labeled with ReAsH. All other proteins were labeled with FlAsH. Scale bars: 2 μm.

We have used both methods to FlAsH-label various yeast strains expressing 2× TetCys-tagged proteins localizing to different cellular structures (amongst others tubulin, actin, mitochondrial inner and outer membrane), demonstrating the general applicability of the method (Figure 4.5). We succeeded in visualizing most FlAsH labeled proteins that could be tagged with the 2× TetCys or the 3× TetCys motif. Nevertheless, image acquisition of less abundant proteins may be challenging, even with sophisticated low-light sensitive microscopes. In our experience, cells labeled with ReAsH generally exhibited an unsatisfactory fluorescence signal so that we abstained from using this compound to visualize proteins expressed at endogenous levels. ReAsH was however used to visualize an overexpressed protein targeted to the mitochondrial matrix (Figure 4.5).

4.1.7 Outlook

The biarsenical-tetracysteine system is an important asset in the toolbox for live cell imaging. Its major advantages are the small size of the tagging motif and the varied (fluorescent) labels that can be utilized for different applications. Nevertheless, at present, fluorescent proteins outperform the biarsenical-tetracysteine system with respect to fluorescence intensity and photostability, albeit, of course, at the cost of a larger tag. Recent progress in the development of novel biarsenical dyes promises to alleviate many of the current limitations, and should make this labeling system even more significant in the future.

4.2
Use of the Biarsenical-Tetracysteine System in *S. cerevisiae*

In this section we focus on two different protocols to label proteins tagged with the TetCys motif expressed in the budding yeast *S. cerevisiae*. Many details of the methods described in the following may be adapted to other organisms.

4.2.1
Materials

4.2.1.1 Growth Media
Growth media exhibiting low autofluorescence are advantageous. Since yeast extract is a major source for autofluorescence in many common growth media, we recommend the use of a synthetic growth medium. The following growth media work well.

Synthetic complete medium (SC-Medium):			
yeast nitrogen base	$1.7\,\mathrm{g\,l^{-1}}$	ammonium sulfate	$5\,\mathrm{g\,l^{-1}}$
complete supplement mixture (see Ref. [31])	$0.6\,\mathrm{g\,l^{-1}}$	glucose	$20\,\mathrm{g\,l^{-1}}$

Mitochondria complete medium:			
yeast nitrogen base	$1.7\,\mathrm{g\,l^{-1}}$	ammonium sulfate	$5\,\mathrm{g\,l^{-1}}$
		galactose	$10\,\mathrm{g\,l^{-1}}$
		ethanol	$20\,\mathrm{ml\,l^{-1}}$
complete supplement mixture (see Ref. [31])	$0.6\,\mathrm{g\,l^{-1}}$	glycerol	$15\,\mathrm{g\,l^{-1}}$

4.2.1.2 Buffers (Required Stock Solutions)

HBS-buffer (pH 7.0):	150 mM NaCl; 20 mM HEPES
Tris solution (pH 7.5):	1 M Tris/HCl

Dye Solution   FlAsH-EDT$_2$ and ReAsH-EDT$_2$ are available from Invitrogen and are delivered in a concentration of 2 mM. Both dyes are sensitive against oxidation and should be aliquoted before use. Freeze and thaw cycles as well as extended light exposures must be avoided.

Biarsenical Dye Working Solution

200 µM FlAsH-EDT$_2$ or 200 µM ReAsH-EDT$_2$ in 1 M Tris-HCl (pH 7.5).

This working solution is needed for the overnight staining protocol only (see below). It should be prepared immediately before use. Storage is not recommended.

Washing Buffer The antidote 1,2-ethanedithiol (Sigma-Aldrich, St.Louis, MO, USA) exhibits a strong offensive smell. A fume hood must be used. We recommend using an extra set of pipettes for EDT because this volatile compound can easily contaminate the pipettes, which may be obstructive for other molecular biology applications.

---

washing buffer: 1 mM EDT in HBS-buffer (vortex thoroughly, prepare freshly)

4.2.2 Labeling Protocols

For the staining of budding yeast cells two different protocols (overnight staining and pulse staining) were established in our laboratory.

For overnight staining, the cells are propagated in a growth medium containing the dye (FlAsH or ReAsH). This approach requires adjustment of the growth parameters and generally results in a uniform labeling of the cell culture. In contrast, pulse staining, which utilizes an electric discharge to trigger the internalization of the dye, is comparatively simple but tends to be of lower efficiency. A detailed description of both methods is given below, followed by a short version of the protocols and a troubleshooting section.

4.2.2.1 Overnight Staining

Here, the cells are grown for 12 h in the dye-containing medium. The long incubation time in the staining medium is necessary to enable the biarsenical dyes to reach the interior of the cells at the concentrations required for sufficient labeling. For this protocol we found the haploid yeast strains of the genetic background BY4741/BY4742 to be more suitable than the diploid strain BY4743.

Preparation of Yeast Cells During overnight staining the cells should be kept continuously in the logarithmic growth phase and therefore the starting cell density has to be carefully adjusted. To this end the growth rate of the yeast strain of interest should be determined in the absence of the biarsenical dye but otherwise under staining conditions. (Since EDT has a slight influence on the doubling time it should be added at this step.) Once the growth rate is determined a logarithmically growing pre-culture is diluted accordingly to proceed with the staining.

FlAsH-Staining The staining of the yeast cells is typically performed in a volume of ~200 µl. To this end, we recommend the use of a 48-well microtiter plate. To avoid excessive evaporation of the growth medium during staining, adjacent wells are filled with media.

For staining 2–6 µl biarsenical dye working solution (final dye concentration: 2–6 µM) and 1 µl washing buffer (final EDT concentration: 5 µM) are added to 200 µl growth medium. Generally, 2–4 µM FlAsH-EDT$_2$ and 4–6 µM ReAsH-EDT$_2$ are sufficient. The EDT is required to lessen unspecific staining. Finally, an adequate amount of cells from a logarithmically growing preculture is added.

The cells are grown overnight at 30 °C in the 48-well microtiter plate in a rotating shaker at ∼120 rpm. To keep them in the dark, the plates are covered with aluminum foil. To maximize the yield, the culture should be grown until the late logarithmic growth phase.

Washing After overnight staining the cells are transferred to an Eppendorf-cup and harvested by centrifugation (2 min, 1500 g). The pellet is resuspended in 1 ml growth medium containing 30–50 µM EDT and incubated for 10–15 min. After this destaining step the cells are peleted and resuspended in growth medium (without EDT). Further washing steps may be beneficial before the cells are mounted for microscopy (see below).

4.2.2.2 Pulse Staining

This approach relies on the electroporation of the cells in the presence of the dye, which is accomplished with a standard electroporator. This method reduces the required time for the whole staining procedure to less than 1 h. Cells grown in liquid media or on agar plates may be used. The unspecific background tends to be considerably lower than in the case of the overnight staining approach. The major disadvantage of this method is the low portion of stained cells. Generally 2–10% of all cells are labeled efficiently.

Preparation of Yeast Cells The cells need not necessarily be in the logarithmic growth phase. However, a large number of dead cells interferes with the staining efficiency because the dead cells tend to accumulate biarsenical dyes.

Prior to electroporation, the cells are washed twice with distilled water (room temperature) to remove ions. The cells are then harvested by centrifugation in an Eppendorf-cup. The amount of cells is not critical. The cell pellet is resuspended in 40 µl distilled water and 0.2–0.4 µl biarsenical dye is added from the undiluted 2 mM stock solution. Finally the cells are transferred in a standard electroporation cuvette with a width of 2 mm.

Staining The uptake of the dye is induced by an electric discharge, delivered by an electroporator. The best staining was achieved with the following parameters: 600 Ω, 0.5 kV, 25 µF. Immediately after electroporation, 160 µl growth medium containing 1 µl fresh washing buffer (final EDT concentration: 5 µM) is added. Before washing the cells are incubated for ∼30 min at 30 °C.

Washing The cells are harvested and destained as described for the overnight procedure (see above). The time needed for destaining may be reduced to ∼5 min.

4.2.2.3 Mounting and Microscopy

Live cells are mounted in growth-medium or buffer solution. To this end the cells are harvested by centrifugation, resuspended in a small volume of growth medium or buffer and mounted below a cover slip. Alternatively, and for longer observation times, the cells may be taken up in 1% low melting agarose (Type VII, Sigma, St Louis,

MO, USA) in growth-medium kept at ~40 °C. However, the low melting agarose may exhibit slight autofluorescence that may interfere with the weak FlAsH signal.

Generally, FlAsH and ReAsH are prone to photobleaching, rendering the imaging potentially challenging. We typically use epifluorescence microscopy to determine the staining efficiency. Subsequently confocal microscopy or other imaging techniques may be used. Imaging with high light intensities should be avoided and sensitive detection systems are advisable.

4.3 Short Protocols

4.3.1 Overnight Staining

Labeling:

- harvest an adequate amount of cells from a logarithmically growing culture (2 min at 1500 g)
- prepare: 200 µl growth-medium with 1 µl fresh washing buffer
- dilute the biarsenical dye 1 : 10 in Tris-buffer
- add 2–6 µl of diluted biarsenical dye to the culture
- inoculate the medium with cells (so that a logarithmic growth for at least 12 h is ensured)
- grow cells for at least 12 h in a rotating shaker at 30 °C (in the dark).

Washing:

- harvest cells by centrifugation in an Eppendorf-cup (2 min at 1500 g)
- resuspend in 1 ml of growth-medium
- add 30–50 µl of fresh washing buffer
- incubate for 10–15 min, pellet the cells by centrifugation (4 min at 1500 g)
- wash the cells for ~1 min in growth-medium to remove EDT
- harvest the cells for microscopy.

Required time: ~13–18 h

4.3.2 Pulse Staining

Labeling:

- prepare: 160 µl growth-medium with 1 µl fresh washing buffer
- wash cells (from liquid culture or agar plate) twice in 1 ml H_2O
- pellet the cells and resuspend them in 40 µl H_2O
- add 0.2–0.4 µl undiluted biarsenical dye to the cells
- transfer the sample into an electroporation cuvette

- Pulse label at 600 Ω, 25 µF, 0.5 kV and immediately add the 160 µl growth-medium with EDT
- transfer the cells to an Eppendorf cup
- incubate in the dark for ∼30 min in a rotating shaker at 30 °C.

Washing:

- harvest cells by centrifugation in an Eppendorf-cup (2 min at 1500 g)
- resuspend in 1 ml growth-medium
- add 30–50 µl of fresh washing buffer
- incubate for ∼5 min, pellet the cells by centrifugation (4 min at 1500 g)
- wash the cells for ∼1 min in growth-medium to remove EDT
- harvest the cells for microscopy.

Required time: ∼1 h

4.4 Troubleshooting

	Problem: Weak or no fluorescence signal
Insufficient labeling	We recommend establishing the labeling protocol with an abundant tagged protein.
	Try the alternative protocol (electroporation or overnight staining).
	EDT may be omitted from the staining solution.
	In rare cases increasing the dye concentration (up to 10 µM) and incubation time (in case of overnight staining) is useful.
Too extensive washing	Reduce the length of the de-staining step and/or reduce the concentration of EDT in the de-staining solution.
The biarsenical dye is not working	Both, FlAsH and ReAsH, are very sensitive to multiple freeze–thaw cycles and oxidation.
	ReAsH changes its color upon oxidation from red to blue/purple. ReAsH with a purple color does not work properly.
	FlAsH and ReAsH aliquots (5–10 µl) are best stored at −80 °C.
Signal is too low	Fusion proteins labeled with the biarsenical-tetracysteine system generally display a weaker fluorescence signal than GFP-fusion proteins.

	The TetCys motif may not be accessible by the dye. The motif might be buried within the folded protein or a membrane.
	The target cysteines must be in their reduced form. Cysteines that are in the lumen of the secretory pathway or outside cells tend to oxidize spontaneously. Use of tributylphosphine to reduce the cysteines [24] may alleviate the problem.
Problem: Background is too high	
Insufficient destaining/ bright vacuoles	Frequently, unbound biarsenical dyes are accumulated in the vacuoles during the staining procedure and a slightly stained cytosol may be observed. This problem is normally alleviated by increasing the duration of the destaining steps and/or the concentration of EDT in the de-staining solution.
Problem: Very bright individual cells	
Dead yeast cells accumulate biarsenical dyes	If too many dead cells are in the solution, this may reduce the amount of free dye and reduce the staining efficiency of the healthy cells. Care must be taken to minimize the number of dead cells.
Problem: Fast loss of fluorescence signal	
Photobleaching	FlAsH and ReAsH are prone to photobleaching. Reduce irradiation intensities.
	At low oxygen concentrations, the overall FlAsH fluorescence signal may decrease. (This happens, for example, when live yeast cells consume the oxygen under the coverslip.) Sometimes a higher FlAsH fluorescence signal is observed in close vicinity to air bubbles.

Acknowledgements

The microphotographs of Figure 4.3 were kindly provided by G.M. Gaietta, R.Y. Tsien, M.H. Ellisman and C. Somerville. We thank Jessica Schilde for providing us with the image of an Mgm1-labeled yeast cell. We thank J. Jethwa for carefully reading the manuscript. S.J. wants to acknowledge financial support by the Deutsche Forschungsgemeinschaft (JA 1129/3).

References

1 Tsien, R.Y. (1998) *Annual Review of Biochemistry,* **67**, 509–544.
2 Shaner, N.C., Campbell, R.E., Steinbach, P.A., Giepmans, B.N., Palmer, A.E. and Tsien, R.Y. (2004) *Nature Biotechnology,* **22**, 1567–1572.
3 Verkhusha, V.V. and Lukyanov, K.A. (2004) *Nature Biotechnology,* **22**, 289–296.
4 Griffin, B.A., Adams, S.R. and Tsien, R.Y. (1998) *Science,* **281**, 269–272.
5 Griffin, B.A., Adams, S.R., Jones, J. and Tsien, R.Y. (2000) *Methods in Enzymology,* **327**, 565–578.
6 Machleidt, T., Robers, M. and Hanson, G.T. (2007) *Methods in Molecular Biology,* **356**, 209–220.
7 Hoffmann, C., Gaietta, G., Bunemann, M., Adams, S.R., Oberdorff-Maass, S., Behr, B., Vilardaga, J.P., Tsien, R.Y., Ellisman, M.H. and Lohse, M.J. (2005) *Nature Methods,* **2**, 171–176.
8 Dyachok, O., Isakov, Y., Sagetorp, J. and Tengholm, A. (2006) *Nature,* **439**, 349–352.
9 Andresen, M., Schmitz-Salue, R. and Jakobs, S. (2004) *Molecular Biology of the Cell,* **15**, 5616–5622.
10 Zhang, J., Campbell, R.E., Ting, A.Y. and Tsien, R.Y. (2002) *Nature Reviews Molecular Cell Biology,* **3**, 906–918.
11 Adams, S.R., Campbell, R.E., Gross, L.A., Martin, B.R., Walkup, G.K., Yao, Y., Llopis, J. and Tsien, R.Y. (2002) *Journal of the American Chemical Society,* **124**, 6063–6076.
12 Martin, B.R., Giepmans, B.N., Adams, S.R. and Tsien, R.Y. (2005) *Nature Biotechnology,* **23**, 1308–1314.
13 Cao, H., Chen, B., Squier, T.C. and Mayer, M.U. (2006) *Chemical Communications,* 2601–2603.
14 Spagnuolo, C.C., Vermeij, R.J. and Jares-Erijman, E.A. (2006) *Journal of the American Chemical Society,* **128**, 12040–12041.
15 Thorn, K.S., Naber, N., Matuska, M., Vale, R.D. and Cooke, R. (2000) *Protein Science,* **9**, 213–217.
16 Gaietta, G., Deerinck, T.J., Adams, S.R., Bouwer, J., Tour, O., Laird, D.W., Sosinsky, G.E., Tsien, R.Y. and Ellisman, M.H. (2002) *Science,* **296**, 503–507.
17 Marek, K.W. and Davis, G.W. (2002) *Neuron,* **36**, 805–813.
18 Tour, O., Meijer, R.M., Zacharias, D.A., Adams, S.R. and Tsien, R.Y. (2003) *Nature Biotechnology,* **21**, 1505–1508.
19 Ju, W., Morishita, W., Tsui, J., Gaietta, G., Deerinck, T.J., Adams, S.R., Garner, C.C., Tsien, R.Y., Ellisman, M.H. and Malenka, R.C. (2004) *Nature Neuroscience,* **7**, 244–253.
20 Nakanishi, J., Maeda, M. and Umezawa, Y. (2004) *Analytical Sciences,* **20**, 273–278.
21 Park, H., Hanson, G.T., Duff, S.R. and Selvin, P.R. (2004) *Journal of Microscopy,* **216**, 199–205.
22 Estevez, J.M. and Somerville, C. (2006) *Biotechniques,* **41**, 569–574.
23 Meeusen, S., DeVay, R., Block, J., Cassidy-Stone, A., Wayson, S., McCaffery, J.M. and Nunnari, J. (2006) *Cell,* **127**, 383–395.
24 Gaietta, G.M., Giepmans, B.N., Deerinck, T.J., Smith, W.B., Ngan, L., Llopis, J., Adams, S.R., Tsien, R.Y. and Ellisman, M.H. (2006) *Proceedings of the National Academy of Sciences of the United States of America,* **103**, 17777–17782.
25 Erster, O., Eisenstein, M. and Liscovitch, M. (2007) *Nature Methods,* **4**, 393–395.
26 Roberti, M.J., Bertoncini, C.W., Klement, R., Jares-Erijman, E.A. and Jovin, T.M. (2007) *Nature Methods,* **4**, 345–351.
27 Tour, O., Adams, S.R., Kerr, R.A., Meijer, R.M., Sejnowski, T.J., Tsien, R.W. and Tsien, R.Y. (2007) *Nature Chemical Biology,* **3**, 423–431.
28 Enninga, J., Mounier, J., Sansonetti, P. and Tran Van Nhieu, G. (2005) *Nature Methods,* **2**, 959–965.
29 De Antoni, A. and Gallwitz, D. (2000) *Gene,* **246**, 179–185.
30 Huh, W.K., Falvo, J.V., Gerke, L.C., Carroll, A.S., Howson, R.W., Weissman, J.S. and O'Shea, E.K. (2003) *Nature,* **425**, 686–691.
31 Sherman, F. (2002) *Methods in Enzymology,* **350**, 3–41.

5
AGT/SNAP-Tag: A Versatile Tag for Covalent Protein Labeling
Arnaud Gautier, Kai Johnsson, and Helen O'Hare

5.1
Introduction

The use of autofluorescent protein fusions to enable microscopic visualization of protein localization, dynamics and function in living cells has revolutionized our understanding of cellular processes. The successes and limitations of autofluorescent proteins have motivated scientists to develop alternative approaches to label proteins with chemical probes in living cells. These techniques combine the convenience of a genetically encoded fusion tag with the power of innovation of organic chemistry. The protein of interest is expressed as a fusion to a polypeptide tag that does not form a fluorophore itself, but reacts specifically with a synthetic fluorophore or other chemical probe. Synthetic fluorophores offer significant advantages over autofluorescent proteins, such as superior brightness, longer fluorescence lifetime, higher photostability or the choice of excitation and emission maxima in the near-infrared region. Synthetic probes can also impart proteins with properties that could not be genetically encoded, for example, environmentally sensitive dyes, photo-crosslinkers or probes for electron microscopy, magnetic resonance imaging or positron emission tomography [1]. Furthermore, since the timing of labeling is under experimental control, protein trafficking, protein turnover, organelle dynamics and macromolecular assembly are open to investigation. Finally, these approaches offer great flexibility, since a single fusion protein can be labeled with any label without the need for re-cloning. This is particularly important when each cloning step represents a considerable investment of effort such as the generation of a stably transfected cell line, or high throughput applications.

In the last few years a number of tag-based techniques have been developed for covalent protein labeling, and commercialized for use outside the laboratory of origin. These include the tetracysteine tag (Invitrogen) [2], O^6-alkylguanine-DNA alkyltransferase (AGT or SNAP-tag, Covalys Biosciences) [3], carrier protein tags (Covalys Biosciences) [4, 5], and the HaloTag (Promega).

These methods have been reviewed previously [6], here we discuss the use of mutants of O^6-alkylguanine-DNA alkyltransferase (AGT or SNAP-tag) as a fusion tag

Probes and Tags to Study Biomolecular Function. Lawrence W. Miller (Ed.)
Copyright © 2008 WILEY-VCH Verlag GmbH & Co. KGaA, Weinheim
ISBN: 978-3-527-31566-6

for protein labeling [3]. Specific labeling of this tag is achieved using O^6-benzylguanine (BG) derivatives, carrying a fluorophore or other label. The SNAP-tag undergoes an irreversible self-labeling reaction, in which the probe is transferred to an active site cysteine to form a stable covalently modified protein. Since the first description of this methodology in 2003, more than a dozen different proteins have been labeled for fluorescence imaging in a variety of subcellular locations of mammalian cell lines and an anaerobic protozoan parasite. The same method has also been used for protein biotinylation, labeling with other small molecules, induced protein dimerization and generation of protein microarrays by self-immobilization. This chapter presents a general overview of the state of the art of the SNAP-tag method, and detailed practical protocols for imaging experiments.

5.2
Labeling SNAP-Tag Fusion Proteins with BG Derivatives

5.2.1
Human O^6-Alkylguanine-DNA Alkyltransferase

Human O^6-alkylguanine-DNA alkyltransferase (hAGT; EC 2.1.1.63) is a DNA repair protein that protects the genome against DNA damage by alkylating agents. Mutagenic alkyl cross-links (adducts) are formed at the O^6 position of the nucleoside guanine by both cellular and exogenous alkylating agents. hAGT removes these alkyl lesions by a stoichiometric one-step "suicide" reaction to restore the native structure of DNA. During the repair process, the O^6-alkyl group is transferred to a reactive cysteine residue in the protein (Cys145) forming a highly stable thioether bond and leading to an irreversibly alkylated protein [7, 8]. The cytotoxicity of O^6-alkylguanine lesions is the basis of anticancer chemotherapies that employ DNA alkylating agents to methylate or chloroethylate the O^6 position of guanine. Consequently, tumor cells expressing high levels of hAGT are resistant to alkylating chemotherapeutic agents, and hAGT itself represents an active anticancer target [7]. An approach to overcome the role of hAGT in therapeutic resistance is to treat patients with alkylating agents and hAGT inhibitors simultaneously. The stoichiometric nature of the reaction facilitates the depletion of hAGT activity using O^6-alkylsubstrates. The most useful such substrate developed so far is O^6-benzylguanine (BG), which inactivates hAGT *in vitro* and *in vivo* [9, 10] and enhances the efficacy of methylating and chloroethylating antitumor drugs in clinical trials [11].

hAGT is a 21 kD monomeric protein of 207 amino acids, with the active site cysteine contained within a consensus sequence PCHR, which is conserved in all known O^6-alkylguanine-DNA alkyltransferases. hAGT is made up of distinct N-terminal and C-terminal lobes (residues 4–91 and 92–176, respectively) [12, 13] (Figure 5.1a). The C-terminal domain contains the active cysteine, the substrate binding pocket and the helix-turn-helix (HTH) DNA-binding motif. The proposed mechanism of hAGT involves a Glu–His–water–Cys hydrogen bond network that probably generates a thiolate anion, explaining the high reactivity of Cys145. In this mechanism, His146 acts as a water-mediated general base to deprotonate Cys145, this acts as the nucleophile in the S_N2 repair reaction. Additionally, the Tyr114 hydroxyl donates a hydrogen bond to N^3 of O^6-alkylguanine, which is thought to promote reactivity by reducing the negative charge on the guanine leaving group [13] (Figure 5.1b).

5.2.2
The Principle of SNAP-Tag Labeling

Recently, mutants of hAGT, hereafter referred to as SNAP-tag, have been used as a genetically encoded tag for specific covalent labeling of fusion proteins *in vitro* and in living cells [3, 14]. This method takes advantage of the high specificity of the reaction of SNAP-tag with O^6-benzylguanine (BG) or BG derivatives with a modified benzyl moiety [15]. Specific labeling in live cells is achieved using cell-permeable BG derivatives carrying chemical probes at the 4-position of the benzyl ring. The derivatized benzyl group is irreversibly transferred to the active site cysteine to produce the labeled protein (Figure 5.1c). Labeling SNAP-tag fusion proteins *in vivo* or *in vitro* comprises three steps: synthesis (or purchase) of a BG derivative carrying the desired molecular probe, expression of the protein of interest as a SNAP-tag fusion, and finally addition of the BG derivative to cells or cell lysate to achieve specific labeling of the fusion protein.

The observation that large synthetic probes can be attached to BG without compromising its activity towards SNAP-tag [3, 14], and the ease with which such BG derivatives can be prepared (Figure 5.2a), make this a versatile and attractive method for a broad range of applications. BG has been derivatized with fluorophores for cell imaging [16] and *in vitro* analysis, and long-life time fluorophores for time-resolved fluorescence resonance energy transfer (FRET) experiments [17] (Figure 5.2b). Spectrally distinct dyes have been developed across the visible spectrum, allowing SNAP-tag labeling to be used in conjunction with other staining procedures or autofluorescent proteins for multicolor imaging (Table 5.1). The availability of multiple BG-fluorophores allows sequential labeling to distinguish newly synthesized protein populations from older generations [18]. BG has also been derivatized with probes for pull-down assays, Western Blot analysis and induction of protein dimerization [19, 20]. Bifunctional probes combining an affinity probe with a fluorophore have been described [21] (Figure 5.2b). The functionalization of various surfaces with BG allows SNAP-tag fusion proteins to be immobilized on chips [22–24] or beads [25]. Many of these probes, as well as functionalized BG derivatives suitable for creating custom labels, are commercially available from Covalys Biosciences.

Figure 5.1 (A) Crystal structure of human O^6-alkylguanine-DNA alkyltransferase (hAGT): the C-terminal lobe contains the active cysteine Cys145, the substrate binding pocket and the HTH DNA-binding motif; (B) the proposed mechanism of hAGT involves a catalytic triad placed to deprotonate the active cysteine Cys145 and a proton-donating residue Tyr114 promoting the dealkylation reaction; (C) general SNAP-tag labeling mechanism: mutant hAGT is used as a genetically encoded tag, known as "SNAP-tag", for specific labeling of the target protein by a O^6-benzylguanine (BG) derivative carrying a chemical probe.

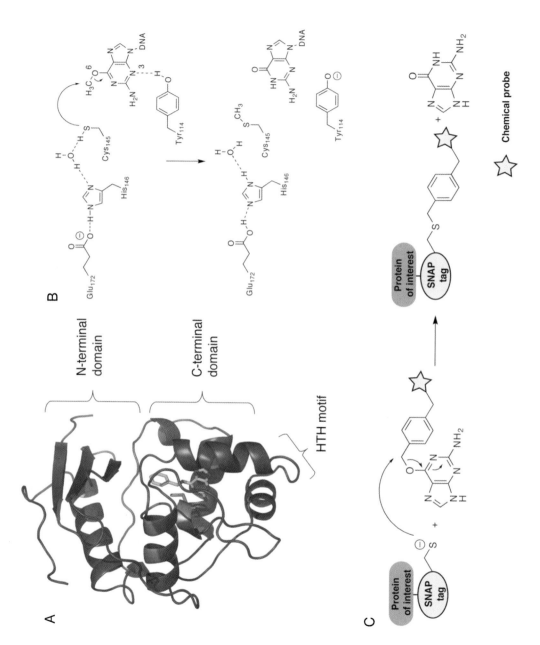

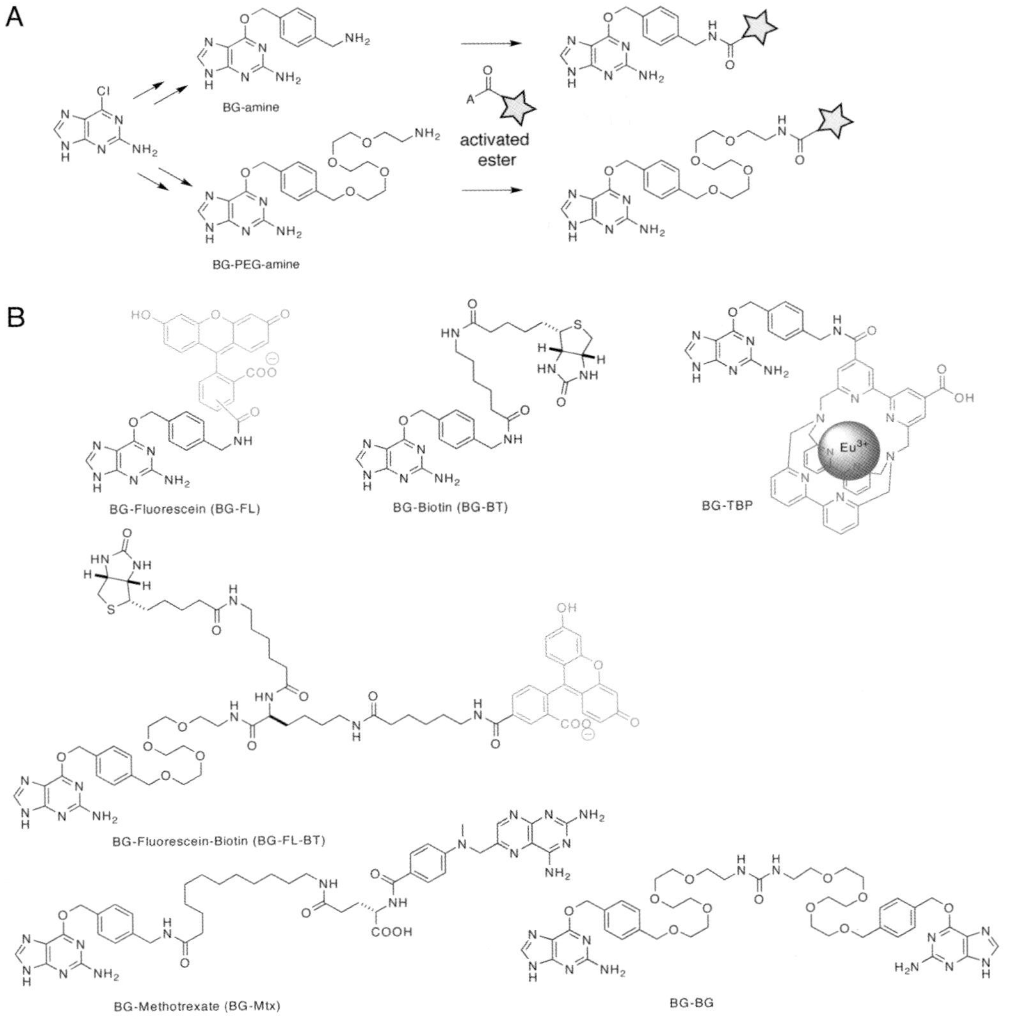

Figure 5.2 (A) BG-amine and BG-PEG-amine are the central precursors for the synthesis of BG derivatives: they can react with virtually any chemical probe available as a free carboxylic acid or activated ester; (B) examples of the functional diversity of available O^6-benzylguanine (BG) substrates. BG has been functionalized with fluorophores (e.g., BG-Fluorescein) for *in vivo* imaging or *in vitro* analysis, with affinity probes (e.g., BG-Biotin) for Western Blot analysis or pull-down assays, and with a long-lifetime fluorophore (BG-TBP) for time-resolved FRET experiments. Bifunctional substrates have been prepared, that combine probes with different properties (e.g., BG-Fluorescein-Biotin). Chemical inducers of dimerization have been synthesized by linking BG to methotrexate (BG-Mtx) or to a second BG molecule (BG–BG). These substrates induce dimerization of the fusion proteins.

Table 5.1 Examples of commercially available SNAP-tag substrates for labeling in living cells.

Substrate	Dye	Abs. (nm)	Em. (nm)	Cell permeable	Ref.
BG-DEAC	diethylaminomethylcoumarin	432	472	yes	[18]
BG-DF	diacetylfluorescein	499	523	yes	[3]
BG-OG	Oregon Green	495	525	yes	[16]
BG-TMR	tetramethylrhodamine	554	580	yes	[16]
TMR-Star[a]	tetramethylrhodamine	554	580	yes	Covalys
BG-Cy3	Cy3	553	572	to be injected	[16]
BG-SF	SNARF	526, 562	621	yes	[16]
BG-Cy5	Cy5	649	670	to be injected	Covalys
BG-TBP	tris-bipyridine Eu(III)	335	580–650	to be injected	Covalys

[a] TMR-*Star* provides the same fluorescent probe as BG-TMR, with the same spectral properties, but is modified at the guanine to increase solubility and cell permeability.

Several properties of hAGT have been optimized by protein engineering and directed evolution to increase the efficiency of protein labeling. The activity towards BG was increased by directed evolution using phage display, the size of the protein was reduced to 182 amino acids by C-terminal truncation, non-essential cysteines were removed to facilitate folding under oxidizing conditions, and mutations were introduced to prevent DNA binding [26–28]. The most active mutant described so far is about 50-fold faster than hAGT, with a second-order rate constant for the reaction with BG of about $10^4 \, M^{-1} \, s^{-1}$ [28].

Concerns over the use of BG derivatives in mammalian cells arose from the potential of endogenous hAGT to become labeled, which might lead to background fluorescence in the nucleus. Such background fluorescence is undetectable in the majority of mammalian cell lines, when labeled according to standard protocols, due to the higher reactivity of SNAP-tag compared to hAGT. Furthermore the concentration of hAGT is often lower than that of the protein of interest. If background is detected, endogenous hAGT can be permanently inactivated by the inhibitor N^9-cyclopentyl-O^6-(4-bromothenyl)guanine (CG), and CG-resistant hAGT mutants can subsequently be labeled with the BG derivative [27] (Figure 5.3a).

More recently, the substrate specificity of hAGT has been reprogrammed by directed evolution to obtain a mutant that reacts with the non-natural substrate O^6-propargylguanine (PG), an extremely poor substrate for wild type hAGT [29] (Figure 5.3b). The specificity was altered by the insertion of a randomized loop in the protein backbone and the mutant, named ^{L}AGT, can be used in conjunction with SNAP-tag for two-color labeling of two different proteins of interest in a single cell.

In summary, the high value of the SNAP-tag lies in the following unique combination of properties:

1. *Efficient, specific and stable labeling* – the speed and high specificity of the reaction of SNAP-tag with otherwise chemically inert BG derivatives allow efficient labeling of the target protein *in vitro* as well as in living cells. The stability of the covalently labeled protein allows experiments of many hours duration and "pulse-chase" labeling. Moreover the labeling reaction is performed with the protein in its native state, but the properties of the added label (or fluorophore) are unaffected by

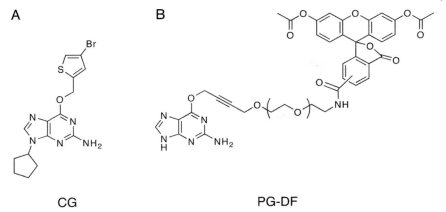

Figure 5.3 Evolution of the substrate recognition of AGT. (A) AGT mutants were selected for resistance to inhibition by N^9-cyclopenthyl-O^6-(4-bromothenyl)guanine (CG). This irreversible inhibitor can be used to inactivate hAGT, whereas the evolved AGT mutants remain active for background-free fluorescent labelling; (B) O^6-propargylguanine (PG) is an extremely poor substrate for SNAP-tag and hAGT, but reacts quickly with evolved ^{L}AGT, allowing dual labeling of SNAP-tag and ^{L}AGT in the same cell.

subsequent modifications, such as cell fixing, protein unfolding for translocation across a membrane, or denaturation for polyacrylamide gel electrophoresis.

2. *A versatile substrate* – Modification of BG derivatives does not affect the reactivity with SNAP-tag, allowing derivatization with a wide range of compounds. The straightforward synthesis of BG compounds facilitates the development of new probes with original properties. Moreover, since labeling procedures do not require a molar excess of label over protein, they make effective use of valuable substrates.

3. *An engineered mutant with improved properties* – SNAP-tag, a monomer of 182 residues, is slightly smaller than autofluorescent proteins (~240 residues), is well expressed in various cell types and is thought to be "localization neutral" so that fusion to this tag should not alter the function or localization of most proteins, though, as for any protein tag, this should be checked experimentally [16]. Additionally, the development of hAGT mutants with altered substrate specificity extends the range of applications of such techniques to allow differential labeling of two protein populations within the same cell [29].

4. *Commercial availability* – The SNAP-tag gene and a large variety of different BG derivatives are available for the biological community from Covalys Biosciences (www.covalys.com).

5.3
SNAP-Tag for Cell Imaging

A number of recent publications demonstrate that SNAP-tag can be used for fluorescent labeling to visualize protein localization in the nucleus, cytoplasm,

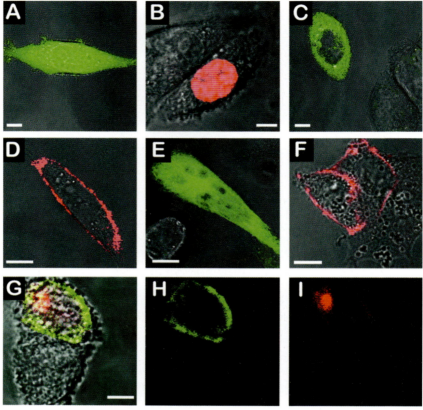

Figure 5.4 Applications of the SNAP-tag in cell imaging. Transiently transfected CHO cells (A–E) or HEK293 cells (F–I), expressing AGT fusion proteins were labeled with BG derivatives. (Scale bars show 10 μm). (A) SNAP-tag labeled using BG-OG; (B) nuclear localized SNAP-NLS labeled using BG-SF; (C) cytoplasmic SNAP-β-galactosidase labeled using BG-DF; (D) SNAP-CaaX localized at the inner surface of the plasma membrane, labeled using BG-TMR; (E) SNAP-tubulin labeled using BG-DF; (F) SNAP-NK1 on the cell surface, labeled using BG-Cy3; (G–I) pulse-chase labeling to follow the change in localization of tsVSVG-SNAP caused by a shift in temperature. Labeling with BG-DF (green) was carried out at 34 °C. Following a shift to 40 °C, newly translated tsVSVG-SNAP was labeled with BG-SF (red). The fluorescence channels for FL (H) and SF (I) are overlaid with the transmission micrograph (G). Copyright 2004 National Academy of Sciences, USA [16].

endoplasmic reticulum and at the cytoplasmic and extracellular face of the plasma membrane (Figure 5.4) [3, 14, 16, 18, 26–30]. The main strengths of this tag compared to autofluorescent proteins reside in the innate advantages of synthetic fluorophores and the ability to control the timing of labeling to investigate protein dynamics. SNAP-tag has been used to visualize the change in intranuclear distribution of the hERα receptor in response to the addition of hormone 4-hydroxytamoxifen [14]. In a recent publication by Regoes *et al.*, SNAP-tag was used for time lapse fluorescence microscopy of nuclear envelope protein RabA in *Giardia intestinalis* [31]. Imaging autofluorescent proteins is problematic in this microaerobic pathogen, since chromophore maturation requires oxygen concentrations that damage or kill these cells. SNAP-tag offers an alternative to autofluorescent proteins when these cannot be in their native form.

The stability of covalent labeling brings the possibility of using different dyes at different time points during the experiment to distinguish between older and newly synthesized proteins in living cells. In these pulse-chase labeling experiments, it is possible to generate distinct populations of otherwise identical proteins whose discriminating features are only determined by the time point of the respective labeling of each population. The first published example of two-color pulse-chase labeling demonstrated the change in localization of a temperature sensitive membrane protein upon shifting to higher temperature [16]. Labeling with the first fluorophore (BG-DF, see Table 5.1) was performed at the permissive temperature, and any remaining unlabeled protein was blocked with non-fluorescent BG. Cells were then allowed to grow at the non-permissive temperature, and newly synthesized protein was labeled with the second fluorophore (BG-SF, Table 5.1). Overlay of fluorescence micrographs shows that protein translated at 34 °C was correctly localized at the membrane, whereas when translation occurred at 40 °C the protein was retained in the endoplasmic reticulum due to misfolding (Figure 5.4).

More recently, Jansen *et al.* took advantage of the principle of pulse-chase labeling to probe the role of centromeric protein A (CENP-A) in centromere determination and cell division. CENP-A is a histone H3 variant that replaces canonical H3 at the centromeres in all eukaryotes, and is central in the localization of centromere and kinetochore components. Pulse-chase labeling of CENP-A-SNAP revealed that protein turnover is slow, and that CENP-A-SNAP is divided equally between sister chromatids during S phase, remaining stably associated with centromeres throughout multiple cell divisions [30]. Quench-pulse labeling was used next to pinpoint the exact phase of the cell cycle at which newly translated CENP-A is deposited on the chromosomes. By applying BG to cells of thymidine arrested at the G_1/S boundary, the cellular pool of CENP-A-SNAP was quenched by modification with the non-fluorescent benzyl group. After removal of BG, cells were released from arrest and allowed to grow until the end of the S phase, during which time a new generation of CENP-A-SNAP was translated. A pulse of TMR-*Star* was applied to label this new generation with tetramethylrhodamine. Assembly into centromeric histones was followed for several hours and was seen to occur uniquely in G1 (Figure 5.5).

5.4
Procedures for SNAP-Tag Labeling

5.4.1
Standard Protocol for Fluorescent Imaging of SNAP-Tagged Proteins in Transiently Transfected Adherent Mammalian Cell Culture

1. Clone the protein of interest as an N- or C-terminal fusion with the SNAP-tag protein in an expression plasmid for mammalian cells (Gateway vectors are available from Covalys Biosciences).

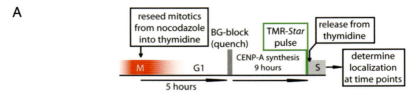

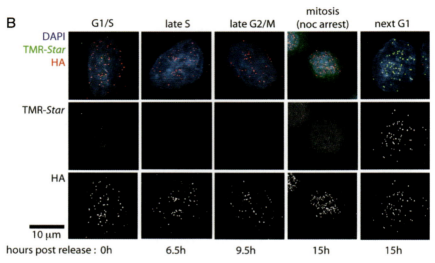

Figure 5.5 Quench-pulse labeling demonstrates cell cycle control of centrosome assembly. Stably transfected HeLa cells expressed CENP A as a fusion to SNAP-tag. A quencher was applied to cells at the start of G1. After 6 h growth, TMR-*Star* was applied to label CENP-A-SNAP synthesized during these 6 h. The entire CENP-A-SNAP population was visualized by immunofluorescence with anti-HA (bottom panel), but only newly synthesized protein was labeled with TMR (middle panel). Reproduced from *Journal of Cell Biology*, 2007, 176: 795–805. Copyright 2007 Rockefeller University Press.

2. SNAP-tag labeling has previously been performed in a large number of common cell lines, including CHO-K1, HEK293, PtK2 and NRK. Prior to experiments seed cells in culture plates appropriate to the microscope (such as Lab TekTM II chambered coverglasses) and grow overnight at 37 °C in an appropriate medium (for example, Dulbecco's modified Eagle medium (DMEM; Invitrogen) without phenol red, and supplemented with 10% fetal calf serum, 10 mM glutamine and antibiotics if required).

3. Transfect cells with the SNAP-tag fusion construct using FuGene (Roche) according to the manufacturer's instructions, or using polyethyleneimine or calcium phosphate according to standard protocols.

4. Incubate cells at 37 °C for 18 to 48 h to allow protein expression. The optimal time depends on the cell line and the protein in question, and should be determined experimentally.

If cells will be cultured further after labeling, use sterile solutions and work within a tissue-culture hood.

5. Dissolve the cell-permeable BG-fluorophore in DMSO at 1 mM. Store this stock at $-20\,°C$ and protect from light. Immediately prior to use, prepare a labeling solution of 5 µM BG-fluorophore in culture medium (or Hank's buffer, or PBS). Prepare the volume that will be used immediately, typically 1 ml per transfection, and mix well. Do not store or reuse labeling medium. The concentration of DMSO in the labeling solution should not exceed 1%. The inclusion of 10% FCS or 1% BSA in the labeling solution may help to prevent background staining.

6. Remove culture medium from transfected cells, add the labeling solution and close the culture plate to prevent evaporation. Incubate at $37\,°C$ for 30 min.

7. Remove labeling medium, wash two times with fresh medium, PBS, or Hank's buffer, then add fresh medium and incubate for 30 min at $37\,°C$. Replace the medium one more time to remove any unreacted label that has leached out of the cells.

8. Image the cells using an appropriate filter set. Most BG-fluorophore labels can be imaged with standard filter sets, for example, BG-DEAC with DAPI filter sets, BG-OG with fluorescein filter sets, TMR-*Star* with rhodamine filter sets.

Tips for optimization of labeling are published online at www.covalys.com. Depending on the particular fluorophore and the protein expression level, the signal to noise ratio can be optimized by varying the concentration of substrate between 1 and 10 µM and varying the labeling period between 10 min and 1 h.

5.4.2
Technical Notes

5.4.2.1 Counterstaining or Fixing
Cells can be counterstained with any live cell dye compatible with the fluorescence properties of the BG-fluorophore. The counterstain can be added to the medium during the 30 min wash (step 7). SNAP-tag labeling is suitable for live cell imaging, but, once labeled, the cells can be fixed by standard procedures using ethanol, methanol or paraformaldehyde, and the fluorescent SNAP-tag fusion proteins visualized as for live cells.

5.4.2.2 Photobleaching
To reduce photobleaching for longer imaging experiments, choose a photostable BG-fluorophore, such as TMR-*Star* or BG-OG, limit exposure of cells to excitation light, use a lower magnification objective or decrease lamp intensity.

5.4.2.3 Checking Expression of the Fusion Protein
Fluorescence imaging of protein gels provides a rapid, sensitive alternative to Western blotting for confirmation of protein expression in mammalian cells.

SNAP-tag fusion proteins can be labeled in cells as above (steps 1–7) or in cell lysates and visualized by fluorescence scanning of polyacrylamide gels. If Western blotting is preferred, detection of the SNAP-tag fusion protein can be achieved using a commercially available monoclonal antibody (Chemicon, mouse anti-MGMT (O^6-methylguanine-DNA methyltransferase) monoclonal antibody, clone MT3.1), or by labeling the SNAP-tag fusion protein with BG-biotin and using appropriate detection reagents.

5.4.2.4 Labeling AGT *in vitro*

To obtain SNAP-tag labeled at almost 100% efficiency, label the pure fusion protein (20 µM) with the required BG derivative (25 µM) in reaction buffer (50 mM Tris-HCl, pH 7.5, 100 mM NaCl, 1 mM DTT). Incubate at ambient temperature for 30 min. Excess label may be removed by standard protein purification procedures (desalting column) if required. Fluorescence scanning of gels is very sensitive (1 fM of labeled protein can be detected), so lower concentrations of protein and BG derivative (1 µM or less) can be used for many applications. To avoid the presence of unreacted label in the final sample, a molar excess of the SNAP-tag fusion protein can be used.

5.4.2.5 Pulse-Chase Labeling

Prepare cells expressing the SNAP-tag fusion protein and perform the first labeling reaction with the first BG-fluorophore, as described above (steps 1–6), then wash twice with medium. During the labeling prepare a stock solution of non-fluorescent inhibitor BG or O^6-(4-bromoethenyl)-guanine by dissolving the inhibitor in DMSO at 1 mM. Prepare a blocking solution by diluting the inhibitor to 10 µM in medium. Remove the medium from the cells, add blocking solution, and incubate for 15 min. This step ensures that 100% of the first SNAP-tag "generation" is either fluorescently labeled or permanently inhibited. After this step only newly synthesized SNAP-tag will be able to react with BG-fluorophores. Wash the cells twice with medium, incubate for 30 min, and replace the medium once more to remove all traces of inhibitor. Incubate the cells in the medium for the required length of time (up to several hours) for growth and new protein synthesis. Labeling with the second BG-fluorophore and imaging are performed according to the standard labeling protocol. A further labeling step with a third BG-fluorophore is also possible.

5.4.2.6 Labeling on the Cell Surface Using Non-Permeable Dyes (BG-FL, BG-Cy3, BG-Cy5)

Non-permeable BG-fluorophores can be used for labeling SNAP-tag fusion proteins on the cell surface. Unlike techniques reliant on autofluorescent proteins, proteins in the secretory pathway will not be labeled, and fluorescent signal will only result from mature proteins localized to the cell surface (subsequent internalization of receptors, however, would lead to fluorescence inside the cell). The labeling protocol is the same as that for cell-permeable BG derivatives, except that the 30 min wash step may be omitted.

5.4.2.7 Labeling Intracellular Proteins Using Non-Permeable Dyes

BG-fluorophores are available across the spectrum, including in the far-red range, where there is no cellular autofluorescence and a lower risk of cell damage due to irradiation. Currently there are no usable far-red-emitting autofluorescent proteins. Dyes in this range offer the possibility of sensitive protein detection, and can add an extra color to a multicolor cell image. Currently all of the far-red-emitting BG-fluorophores are cell-impermeable, however these fluorophores have been used successfully to image intracellular proteins by microinjection of the dye [18] or labeling performed after cell fixing and permeabilization (www.covalys.com). For microinjection, dyes should be diluted in water to 30 µM and injected using an injection capillary micropipet, applying 50 hPa pressure for 0.8 s (InjectMan NI 2 micromanipulator, Eppendorf).

5.4.2.8 Multicolor Labeling of More Than One Protein

To enable labeling of two different proteins with different fluorophores in the same cell, a mutant of AGT, ^{L}AGT has been developed with altered recognition of the alkyl group. ^{L}AGT reacts with O^6-propargylguanine (PG) fluorophores, whereas SNAP-tag does not react with these molecules. Since ^{L}AGT retains activity towards BG-fluorophores, the order of addition of the labeling reagents is crucial. Prepare cells co-expressing an ^{L}AGT fusion protein and a SNAP-tag fusion protein. First, add medium containing 20 µM PGDF to fully label ^{L}AGT fusion proteins with fluorescein. Next, add a labeling solution containing 5 µM of a BG-fluorophore (such as TMR-Star), incubate for 30 min, and continue with the standard protocol for washing and imaging.

5.4.2.9 Measuring Protein–Protein Interactions or Conformational Changes by FRET

The SNAP-tag labeling technology has a particular advantage for studying homo-oligomerization in living cells. Two BG-fluorophores (a donor and acceptor) can be added simultaneously to label the protein population. In order to label 50% of the protein with the first fluorophore and 50% with the second it is sufficient to optimize the ratio of the BG–substrate concentrations (by comparison, the control of the expression level of two different autofluorescent fusion proteins is technically difficult). Furthermore, different BG derivatives can be tested to find the best FRET pair for a particular application without the need to re-clone the protein of interest. In order to study hetero-oligomerization or conformational changes (when two fluorophores are present in a single molecule), two tags with orthogonal substrate specificity are required. This application was demonstrated by Heinis et al. [29] using SNAP-tag and ^{L}AGT. Alternatively, the SNAP-tag can be used in conjunction with another tag such as an autofluorescent protein. SNAP-EGFP labeled with BG-SF led to extremely efficient FRET, since SF has a broad excitation range and is therefore a good FRET acceptor [16]. Time-resolved FRET using a long-lifetime fluorescence donor offers exquisite sensitivity compared to standard dyes or fluorescent proteins. One FRET pair for time resolved FRET of SNAP-tag fusion proteins is commercially available: the BG-TBP Europium cryptate donor and the acceptor BG-Cy5. Since neither molecule is cell-permeable, this pair must be microinjected, used at the cell surface, or added to cell lysates. Considering the broad range of BG-fluorophores available, there is the intriguing possibility that two independent FRET experiments might be performed simultaneously in the same cell.

5.4.2.10 Sensors

Unlike the fluorophores of autofluorescent proteins, which are predominantly shielded from their environment, certain chemical fluorophores show great sensitivity to pH or the concentration of ions such as calcium. The possibility of coupling such fluorophores to benzylguanine for reaction with SNAP-tag at a defined subcellular location offers great promise for high-resolution probes of the cellular microenvironment.

5.5
Broader Applications of SNAP-Tag to Study Protein Function

The SNAP-tag labeling method has been shown to be a general tool for live cell imaging of proteins. Apart from imaging applications, the versatility of the SNAP-tag has allowed the development of many new tools to investigate and manipulate proteins in a more general context, most notably by (i) *in vivo* induction of protein dimerization and (ii) covalent immobilization of SNAP-tag fusion proteins.

5.5.1
Induction of Protein Dimerization by Covalent Labeling in Living Cells

Chemical inducers of dimerization (CIDs) have been introduced to study processes involving protein dimerization or protein–protein interactions in general. CIDs are small molecules that can be used to induce protein dimerization and study the role of protein complexes in biological processes such as signal transduction, transcription, apoptosis, protein degradation and localization [32]. CIDs function by binding two different proteins simultaneously, leading to their dimerization. Commonly used CIDs rely on non-covalent interactions. BG-based CIDs were introduced in order to induce covalent dimerization using the high selectivity of the SNAP-tag labeling.

The first example of a BG-based CID described in the literature was BG-Mtx, in which BG is coupled to methotrexate (Mtx), a tight-binding inhibitor of dihydrofolate reductase (DHFR) (see Figure 5.2b for the structure of BG-Mtx). BG-Mtx was used to control transcription in yeast via a "three-hybrid system" [19]. In a proof-of-principle experiment, the DNA-binding domain LexA expressed as a SNAP-tag fusion, and the transcription activation domain B12 was fused to DHFR. In the presence of BG-Mtx, dimerization of the SNAP-tag and DHFR fusion proteins led to expression of the reporter genes *HIS3* and *lacZ*, allowing growth of the auxotrophic yeast strain on histidine deficient media (Figure 5.6a). The level of β-galactosidase activity resulting from induced dimerization of SNAP-tag and DHFR was shown to be higher than that resulting from dimerization of LexA-Fos and B42-Jun, a protein pair known to yield a strong interaction signal in the two-hybrid system [19].

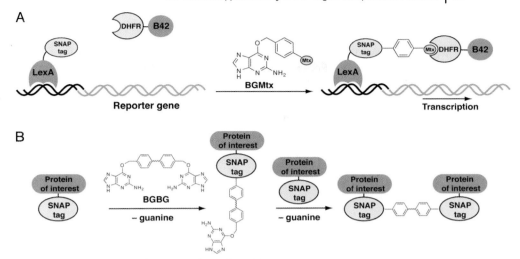

Figure 5.6 Covalent protein dimerization by BG-based CIDs: (A) A SNAP-tag-based three-hybrid system using BG-Mtx-induced dimerization of SNAP-tag and DHFR to activate transcription in yeast; (B) covalent and irreversible dimerization of SNAP-tagged proteins with the BG-BG dimer can be used to probe spatial proximity or detect protein–protein interaction.

This SNAP-tag-based three-hybrid system has been used to isolate hAGT mutants with enhanced activity towards BG. Since the growth rate on selective plates depends critically on the rate of reaction between the SNAP-tag and BG-Mtx, it was possible to isolate improved hAGT mutants from randomized libraries by a simple growth assay [28].

BG-Mtx achieves dimerization by the covalent reaction of SNAP-tag with the BG moiety and the tight binding but non-covalent interaction of Mtx with DHFR. More recently, CIDs have been designed with two BG subunits connected by a flexible linker to allow induction of fully covalent dimerization (for structure example see Figure 5.2b) [20]. Covalent dimerization offers significant advantages over non-covalent interactions. The extent of dimerization of two proteins can easily be quantified by Western Blot analysis under denaturing conditions, thanks to the stability of the covalent crosslink. The extent of protein dimerization could thus be correlated with a biological phenotype. It is also possible to measure the spatial proximity of two proteins and to detect protein–protein interactions via this method, since the efficiency of BG–BG induced cross-linking of the two SNAP-tagged proteins depends on their spatial proximity in the cell [20] (Figure 5.6b). This application was demonstrated by the addition of BG–BG to HEK293 cells co-expressing SNAP-βGal (SNAP-tagged β-galactosidase) and SNAP-EGFP (SNAP-tagged enhanced green fluorescent protein). As well as homodimers of each protein, BG–BG also induced heterodimer formation, since both proteins were resident in the cytosol. By comparison, no heterodimer was detected when the experiment was repeated using nuclear localized SNAP-EGFP-NLS and cytoplasmic SNAP-βGal.

The ratio of homodimer to heterodimer allows the detection of protein–protein interaction, since interactions favor heterodimer formation. This was demonstrated for the rapamycin-dependent interaction between FK506-binding protein (FKBP) and the FKBP-rapamycin associated protein binding domain (FRB). The presence of rapamycin led to an 11-fold increase in the efficiency of BG–BG crosslinking of SNAP-EGFP-FKBP and SNAP-FRB.

In summary, BG-based CIDs can be used to induce covalent dimerization proteins in a variety of different organisms. The specificity of the labeling and its independence of the nature of the ligand should make the approach a valuable tool to control protein dimerization *in vivo* for various applications such as screening of binding partners or protein localization.

5.5.2
SNAP-tag-Mediated Covalent Immobilization of Fusion Proteins on BG-Functionalized Surfaces

The development of new strategies to attach recombinant proteins to solid surfaces in a selective, irreversible and defined manner has emerged as an important challenge in biotechnology, particularly in functional proteomics. Protein function microarrays, microbeads, and protein sensor chips have the potential to become key research tools for analysis of protein function. Multiple strategies have been devised for covalent attachment of proteins of interest by reaction with electrophilic groups such as aldehydes or activated esters displayed on the surface. The main drawback of such strategies is the low specificity of these reactive groups: proteins must be purified before immobilization, a major drawback for high-throughput applications. The SNAP-tag can be used for covalent and selective immobilization of tagged proteins onto BG-derivatized surfaces (Figure 5.7). SNAP-tag-mediated immobilization offers the following advantages: (i) immobilization occurs exclusively via the SNAP-tag and leaves the protein of interest accessible for interaction with other molecules, (ii) attachment is covalent, ensuring that proteins remain linked to the solid phase under a variety of different conditions, (iii) functionalization is highly chemoselective and can be carried out directly from crude cell lysates without the need for time-consuming purification steps, and (iv) attachment is carried out under physiological conditions, preserving the function of the protein of interest.

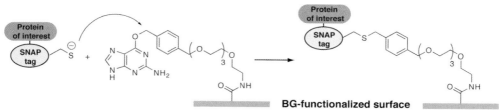

Figure 5.7 SNAP-tag-mediated covalent immobilization of fusion proteins on BG-functionalized surfaces.

BG-activated sensor chips have been prepared by coupling a BG derivative to a commercially available carboxymethylated dextran gold chip (Biacore) suitable for surface plasmon resonance (SPR) experiments. Protein immobilization on these chips was extremely efficient, whether the SNAP-tagged protein was purified first, or immobilized directly from *E. coli* cell lysates. This was illustrated by selective immobilization of GST-SNAP fusion proteins and their subsequent detection using an anti-GST antibody [22]. Huber and Schlatter and coworkers demonstrated that this method is suitable for binding studies with small molecules [33]. They immobilized SNAP-tagged cyclophilin D on a BG-activated SPR sensor chip, and monitored the binding of the ligand cyclosporin A, showing a full retention of cyclophilin D activity after the immobilization step.

A system based on SNAP-tag fusion proteins has been introduced to generate protein function microarrays [24]. SNAP-tagged proteins can be covalently immobilized on glass slides displaying BG. The "bioinertia" of the glass slides was improved by coating them with polymer brushes of poly(oligoethyleneglycol)methacrylate (POEGMA), lowering unspecific binding [23]. The SNAP-tag is attractive for applications on protein function microarrays for two main reasons: (i) the specificity of the reaction between SNAP-tag and BG that allows direct immobilization from a complex mixture, and (ii) the potential to use a single versatile tag for immobilization and for fluorescence-labeling of proteins, which would facilitate high-throughput screening of protein–protein interactions. The same SNAP-tag constructs could be used to generate a defined array of SNAP-tagged proteins and also to generate fluorescently-labeled proteins to probe this array. SNAP-tag-based protein microarrays have been demonstrated to be useful tools for the detection of protein–protein or small molecule–protein interactions and the analysis of post-translational modifications [24].

5.6
Conclusion

In conclusion, the SNAP-tag is a valuable addition to the armory of the cell biologist, which brings a new dimension to well-established methods of imaging autofluorescent proteins in live cells. SNAP-tag labeling is a fast, convenient and specific way to add an extra color to multi-color imaging experiments. The ability to change the fluorophore without changing the underlying genetic construct brings unprecedented flexibility of experimental design. This latter point can be exploited for temporal control of protein labeling, to address previously intractable questions concerning protein dynamics. In addition to cell imaging the tag is advantageous for studying protein function *in vitro*, including high throughput experiments. Recently published applications of the SNAP-tag have demonstrated its potential to tackle fundamental biological questions in living cells, and increased usage of this tag will, in turn, drive further technical innovations for more challenging applications, such as imaging whole organisms.

References

1 Johnsson, N. and Johnsson, K. (2007) Chemical tools for biomolecular imaging. *ACS Chemical Biology*, **2**, 31–38.
2 Griffin, B.A., Adams, S.R. and Tsien, R.Y. (1998) Specific covalent labeling of recombinant protein molecules inside live cells. *Science*, **281**, 269–272.
3 Keppler, A., Gendreizig, S., Gronemeyer, T., Pick, H., Vogel, H. and Johnsson, K. (2003) A general method for the covalent labeling of fusion proteins with small molecules in vivo. *Nature Biotechnology*, **21**, 86–89.
4 Yin, J., Liu, F. and Walsh, C.T. (2004) Labeling proteins with small molecules by site-specific postranslationnal modification. *Journal of the American Chemical Society*, **126**, 7754–7755.
5 George, N., Pick, H., Vogel, H., Johnsson, N. and Johnsson, K. (2004) Specific labeling of cell surface proteins with chemically diverse compounds. *Journal of the American Chemical Society*, **126**, 8896–8897.
6 Marks, K.M. and Nolan, G.P. (2006) Chemical labeling strategies for cell biology. *Nature Methods*, **3**, 591–596.
7 Pegg, A.E., Dolan, M.E. and Moschel, R.C. (1995) Structure, function and inhibition of O^6-alkylguanine-DNA alkyltransferase. *Progress in Nucleic Acid Research and Molecular Biology*, **51**, 167–223.
8 Pegg, A.E. (2000) Repair of O^6-alkylguanine by alkyltransferase. *Mutation Research*, **462**, 83–100.
9 Dolan, M.E., Moschel, R.C. and Pegg, A.E. (1990) Depletion of mammalian O^6-alkylguanine-DNA alkyltransferase activity by O^6-benzylguanine provides a means to evaluate the role of this protein in protection against carcinogenic and therapeutic alkylating agents. *Proceedings of the National Academy of Sciences*, **87**, 5368–5372.
10 Pegg, A.E., Boosalis, M., Samson, L., Moschel, R.C., Byers, T.L., Swenn, K. and Dolan, M.E. (1993) Mechanism of inactivation of human O^6-alkylguanine-DNA alkyltransferase by O^6-benzylguanine. *Biochemistry*, **32**, 11998–12006.
11 Rabik, C.A., Njoku, M.C. and Dolan, M.E. (2006) Inactivation of O^6-alkylguanine-DNA alkyltransferase as a means to enhance chemotherapy. *Cancer Treatment and Research*, **32**, 261–276.
12 Wibley, J.E.A., Pegg, A.E. and Moody, P.C.E. (2000) Crystal structure of the human O^6-alkylguanine-DNA alkyltransferase. *Nucleic Acids Research*, **28**, 393–401.
13 Daniels, D.S., Woo, T.T., Luu, K.K., Noll, D.M., Clarke, N.D., Pegg, A.E. and Trainer, J.A. (2004) DNA binding and nucleotide flipping by the human DNA repair protein AGT. *Nature Structural & Molecular Biology*, **11**, 714–720.
14 Keppler, A., Kindermann, M., Gendreizig, S., Pick, H., Vogel, H. and Johnsson, K. (2004) Labeling of fusion proteins of O^6-alkylguanine-DNA alkyltransferase with small molecules in vivo and in vitro. *Methods*, **32**, 437–444.
15 Damoiseaux, R., Keppler, A. and Johnsson, K. (2001) Synthesis and applications of chemical probes for human O^6-alkylguanine-DNA alkyltransferase. *ChemBioChem*, **2**, 285–287.
16 Keppler, A., Pick, H., Arrivoli, C., Vogel, H. and Johnsson, K. (2004) Labeling of fusion proteins with synthetic fluorophores in live cells. *Proceedings of the National Academy of Sciences*, **101**, 9955–9959.
17 SNAP-vitro HTRF is a BG-substrate incorporating a long-lifetime fluorophore commercialized by Covalys Biosciences and developed with Cisbio.
18 Keppler, A., Arrivoli, C., Sironi, L. and Ellenberg, J. (2006) Fluorophores for live cell imaging of AGT fusion proteins across the visible spectrum. *Biotechniques*, **41**, 167–174.
19 Gendreizig, S., Kindermann, M. and Johnsson, K. (2003) Induced protein

dimerization *in vivo* through covalent labeling. *Journal of the American Chemical Society*, **125**, 14970–14971.
20. Lemercier, G., Gendreizig, S., Kindermann, M. and Johnsson, K. (2007) Inducing and sensing protein–protein interactions in living cells via selective crosslinking. *Angewandte Chemie-International Edition*, **46**, 4281–4284.
21. Kindermann, M., Sielaff, I. and Johnsson, K. (2004) Synthesis and characterization of bifunctional probes for the specific labeling of fusion proteins. *Bioorganic & Medicinal Chemistry Letters*, **14**, 2725–2728.
22. Kindermann, M., George, N., Johnsson, N. and Johnsson, K. (2003) Covalent and selective immobilization of fusion proteins. *Journal of the American Chemical Society*, **125**, 7810–7811.
23. Tugulu, S., Arnold, A., Sielaff, I., Johnsson, K. and Klok, H.A. (2005) Protein-functionalized polymer brushes. *Biomacromolecules*, **6**, 1602–1607.
24. Sielaff, I., Arnold, A., Godin, G., Tugulu, S., Klok, H.-A. and Johnsson, K. (2006) Protein function microarrays based on self-immobilizing and self-labeling fusion proteins. *ChemBioChem*, **7**, 194–202.
25. BG-derivatized beads are produced by Covalys Bioscience under the name SNAP-capture.
26. Juillerat, A., Gronemeyer, T., Keppler, A., Gendreizig, S., Pick, H., Vogel, H. and Johnsson, K. (2003) Directed evolution of O^6-alkylguanine-DNA alkyltransferase for efficient labeling of fusion proteins with small molecules *in vivo*. *Chemistry & Biology*, **10**, 313–317.
27. Juillerat, A., Heinis, C., Sielaff, I., Barnikow, J., Jaccard, H., Kunz, B., Terskikh, A. and Johnsson, K. (2005) Engineering substrate specificity of O^6-alkylguanine-DNA alkyltransferase for specific protein labeling in living cells. *ChemBioChem*, **6**, 1263–1269.
28. Gronemeyer, T., Chidley, C., Juillerat, A., Heinis, C. and Johnsson, K. (2006) Directed evolution of O^6-alkylguanine-DNA alkyltransferase for applications in protein labeling. *Protein Engineering Design and Selection*, **19**, 309–316.
29. Heinis, C., Schmitt, S., Kindermann, M., Godin, G. and Johnsson, K. (2006) Evolving the substrate specificity of O^6-alkylguanine-DNA alkyltransferase through loop insertion for applications in molecular imaging. *ACS Chemical Biology*, **1**, 575–589.
30. Jansen, L.E.T., Black, B.E., Foltz, D.R. and Cleveland, D.W. (2007) Propagation of centromeric chromatin requires exit from mitosis. *Journal of Cell Biology*, **176**, 795–805.
31. Regoes, A. and Hehl, A.B. (2005) SNAP-tag mediated live cell labeling as an alternative to GFP in anaerobic organisms. *Biotechniques*, **39**, 809–812.
32. Klemm, J.D., Schreiber, S.L. and Crabtree, G.R. (1998) Dimerization as a regulatory mechanism in signal transduction. *Annual Review of Immunology*, **16**, 569–592.
33. Huber, W., Perspicace, S., Kohler, J., Müller, F. and Schlatter, D. (2004) SPR-based interaction studies with small molecular weight ligands using hAGT fusion proteins. *Analytical Biochemistry*, **333**, 280–288.

6
Trimethoprim Derivatives for Labeling Dihydrofolate Reductase Fusion Proteins in Living Mammalian Cells
Lawrence W. Miller and Virginia W. Cornish

6.1
Introduction

For over 50 years, biochemists and biophysicists have labeled proteins with fluorescent probes, affinity reagents and radioactive tags in order to interrogate their structure and function *in vitro* [1]. Indeed, a variety of chemical techniques have been developed to label specific amino acid residues of purified proteins in aqueous solutions [2]. When coupled with traditional biochemical methods, these bioconjugation chemistries are immensely useful for elucidating protein structure and dynamic conformational change. However, a complete, mechanistic understanding of a protein's biological function requires the ability to analyze the dynamic distribution, transient interactions, and chemical environment of proteins in living cells. Traditional methods of protein labeling are often inadequate for *in vivo* studies because they require purification of the protein, chemical labeling, repurification and reintroduction into cells by invasive methods such as microinjection. Furthermore, these onerous methods are limited to the subset of proteins that can be overexpressed and purified, precluding their application to the study of most membrane proteins. The limitations of traditional bioconjugation techniques have spawned efforts to non-invasively and site-specifically label proteins in living cells or tissue.

The most prominent method of protein labeling is to genetically encode Green Fluorescent Protein (GFP) or one of its variants as a fusion to the protein of interest [3, 4]. The resulting gene fusion is expressed, and the autofluorescent GFP fusion is detected microscopically. The relatively small size (about 27 kDa) of GFP and its compact, single-domain structure allow it to be fused to a wide variety of target proteins with little or no interference in native protein functionality. A drawback to GFPs, however, is that their spectral and structural characteristics are interdependent [3]. While mutagenesis has led to the development of differently colored GFPs, including cyan, green, yellow and blue variants, and a red-emitting protein dubbed DsRed has been cloned from Discosoma [5–7], it has been difficult to engineer GFP variants with well resolved absorption and emission spectra for

Probes and Tags to Study Biomolecular Function. Lawrence W. Miller (Ed.)
Copyright © 2008 WILEY-VCH Verlag GmbH & Co. KGaA, Weinheim
ISBN: 978-3-527-31566-6

multi-color co-localization and fluorescence resonance energy transfer (FRET) applications, and in particular to obtain a well behaved red variant. In order to increase the diversity of protein labels, approaches are needed that combine the ability to genetically encode the label as for GFP with the functional variability of small molecule labels.

Beginning with the report of the fluorescein-based bis-arsenical dye, FlAsH in 1998 [8], chemists have sought ways to link small molecules to proteins *in vivo* with the same selectivity and facility that is available with GFP [9–11]. The general strategy of *in vivo*, site-specific protein labeling entails genetically fusing a target protein to a receptor protein, protein domain or peptide sequence (Figure 6.1). The small molecule probe consists of a receptor-binding ligand coupled to a fluorophore or other functional moiety. When added to cells growing in culture, the probe binds specifically and stably to the receptor fusion. The success of this strategy depends on identifying or developing a receptor that is specific for the small molecule and that does not interfere with the function of the target protein. For labeling intracellular proteins, the small molecule probe needs to be cell-membrane-permeable. Besides FlAsH, which binds to a tetra-cysteine peptide, other ligand–receptor pairs that have been used for intracellular protein labeling include benzyl guanine and human alkylguanine alkyl transferase [12], methotrexate and eDHFR [13], and synthetic ligation factor and FKBP12 (F36V) [14]. Both acyl carrier protein and peptide carrier protein have been used to label phosphopantetheine transferase with membrane-impermeable, coenzymeA derivatives on cell surfaces [15, 16]. Cell surface proteins fused to a 15-residue acceptor peptide sequence can be labeled with biotin derivatives via a biotin ligase-mediated process [17, 18].

Our strategy for selective, *in vivo* protein labeling makes use of the strong, non-covalent interaction between *Escherichia coli* dihydrofolate reductase (eDHFR) and 2,4-diamino-5-(3,4,5-trimethoxybenzyl) pyrimidine, or trimethoprim (TMP) [19]. The TMP-eDHFR system has a number of qualities that allow for facile protein labeling in living eukaryotic cells. First, eDHFR is a single-domain, monomeric protein, and its small size (18 kDa) compares favorably with GFP. These features suggest that eDHFR can be appended to a wide variety of eukaryotic proteins without compromising native function. Secondly, TMP binds to eDHFR with far greater affinity (K_D ca. 10^{-11} M) than to mammalian forms of the enzyme (K_D ca. 10^{-6} M for binding to murine L1210 DHFR) [20]. This selectivity for eDHFR makes it possible to label the cytosolic face of plasma membrane proteins, as well as cytosolic proteins and proteins associated with internal membranes. These experiments can be carried out in wild-type eukaryotic cells without undue toxicity or high background from binding to endogenous DHFR.

We have recently developed membrane-permeable, fluorescent TMP derivatives that retain high affinity for eDHFR (K_D ca. 10^{-8} M). These derivatives rapidly (<1 h) achieve steady-state concentration conditions when added to cell growth medium at micromolar concentrations [21]. The TMP-linked fluorophores diffuse freely within wild-type mammalian cells, and they do not bind non-specifically to endogenous receptors. The use of these molecules allows fluorescence microscopic detection of the subcellular localization of labeled eDHFR fusion proteins with high signal-to-noise

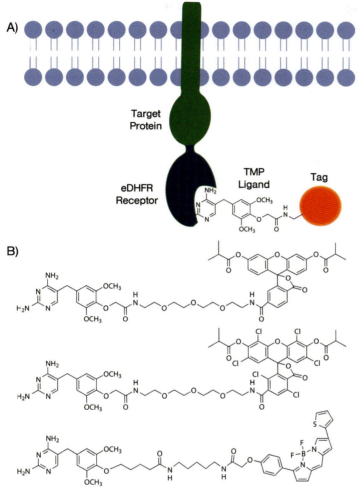

Figure 6.1 (A) Schematic of intracellular protein labeling. Mammalian cells expressing a target protein fused to *E. coli* dihydrofolate reductase (eDHFR) receptor are incubated in growth medium containing a fluorescent trimethoprim (TMP) derivative. The tagged TMP derivative enters the cells and binds selectively to the eDHFR receptor. After washing excess dye from the growth medium, the fluorescently labeled target protein can be visualized microscopically; (B) Structures of fluorescent TMP derivatives described in this chapter. Top: LigandLink fluorescein (Active Motif, Inc., cat. no. 34 101); Middle: LigandLink hexachlorofluorescein (Active Motif, Inc., cat. no. 34 104); Bottom: TMP-BODIPY Texas Red.

ratio. Here we detail methods for expressing eDHFR fusion proteins in mammalian cell lines, labeling them with TMP-linked fluorophores, and analyzing the cells microscopically. We also provide conditions for synthesizing TMP-linked probe molecules.

6.2
Preparation of *E. coli* Expression Vectors

6.2.1
Materials

Restriction enzymes, Vent DNA polymerase and T4 DNA ligase were purchased from New England Biolabs. The dNTPs used in the polymerase chain reaction (PCR) were purchased from Amersham Biosciences. Oligonucleotides were purchased from Invitrogen. The pMONDHFR plasmid was obtained from Professor Jim Hu. The pECFP-N1-lyn plasmid was obtained from Professor Tobias Meyer. The pECFP-NUC plasmid was obtained from Professor Kai Johnsson. The MLCK-DHFR plasmid was prepared by Gregory Giannone in Professor Michael Sheetz' laboratory. Plasmid pLL-1 is available for purchase from Active Motif, Inc.

6.2.2
Plasmids for Over-Expression of *E. coli* DHFR Fusion Proteins in Mammalian Cells

Plasmid vectors for the constitutive over-expression of eDHFR fusion proteins in mammalian cells under control of a cytomegalovirus (CMV) promoter were prepared using standard molecular biology techniques [22]. The vectors included elements for selection of stable transfectants using G418 as well as selection and propagation in *E. coli*. The eDHFR fusion genes begin with a Kozak consensus translation initiation site. DNA encoding various proteins or peptide signal sequences was inserted in the frame at the C- or N- terminus of the eDHFR DNA with an intervening linker sequence (at minimum, Glycine-Serine-Glycine) as warranted. A vector that allows facile preparation of C-terminal eDHFR fusions to a protein or peptide of interest (POI :: eDHFR) is available for purchase from Active Motif, Inc. (cat. no. 34 001, www.activemotif.com).

6.2.2.1 Nucleus-Localized eDHFR Expression Plasmid (pLM1302). Construction of Nucleus-Localized eDHFR Vector (Plasmid pLM1302)

The plasmid pLM1302 was prepared by replacing the gene for CFP in pECFP-NUC (Clontech) with DNA encoding eDHFR. A 521 bp NheI to XhoI fragment encoding DHFR with an N-terminal Kozak sequence and a valine in the second position was prepared by PCR from pMONDHFR using the primers 5′-GCA TAC GTC GCT AGC GCT ACC GGT CGC CAC CAT GGT GAT CAG TCT GAT TGC GGC-3′ (NheI, coding strand) and 5′-GCA TAC GTC CTC GAG ATC TGA GTC CGG ACC GCC GCT CCA GAA TC-3′ (XhoI, non-coding strand). This fragment was inserted between the NheI site and the XhoI site in pECFP-NUC to give to pLM1302. Upon transfection into mammalian cells, pLM1302 expresses the protein fusion eDHFR-SGLRSRA-(DPKKKRKV)$_3$-GSTGS.

6.2.2.2 Myosin Light Chain Kinase eDHFR Plasmid

The gene encoding eDHFR was subcloned from plasmid pMONDHFR to MLCK-GFP to generate MLCK-DHFR. The plasmid MLCK-GFP was prepared by subcloning the

DNA for the full length avian form of MLCK into the *Bam*H1/*Eco*R1 sites of pEGFP-N1(Clontech) [23]. A 518 bp *Bam*H1 to *Not*1 fragment encoding eDHFR was prepared by PCR from pMONDHFR using the primers 5′-GGA TCC TGG AAT GAT CAG TCT GAT TGC GGC GTT AG (*Bam*H1, coding strand) and 3′-GCG GCC GCT TAC CGC CGC TCC AGA ATC TC-3′ (*Not*1, non-coding strand). This fragment was inserted between the *Bam*H1 site and the *Not*1 site in MLCK-GFP to give MLCK-eDHFR. Upon transfection into mammalian cells, MLCK-eDHFR expresses the protein fusion MLCK-GDPGM-eDHFR.

6.3
Synthesis of Fluorescent Trimethoprim Derivatives

The first proof-of-principle experiments that validated the TMP/eDHFR protein labeling method were performed with TMP-BODIPY Texas Red and TMP-fluorescein derivatives [19]. TMP was substituted at the 4′-methoxy position with a straight-chain alkane linker terminated with a carboxylic acid. The linker served to spatially separate the fluorophore from the TMP moiety, and the carboxyl group served as a reactive handle for coupling to commercially available cadaverine derivatives of fluorescein or BODIPY Texas Red (Figure 6.2). Briefly, TMP was converted to 2,4-diamino-5-(3,5-dimethoxy, 4-hydroxybenzyl) pyrimidine **2** by preferential cleavage of the 4′-methoxy group in 48% HBr. Bromoalkylation of the phenol with ethyl 5-bromo-valerate in DMSO with potassium *tert*-butoxide yielded compound **3**. Compound **3** was hydrolyzed in methanol with NaOH, purified by crystallization, and coupled to the respective fluorophores.

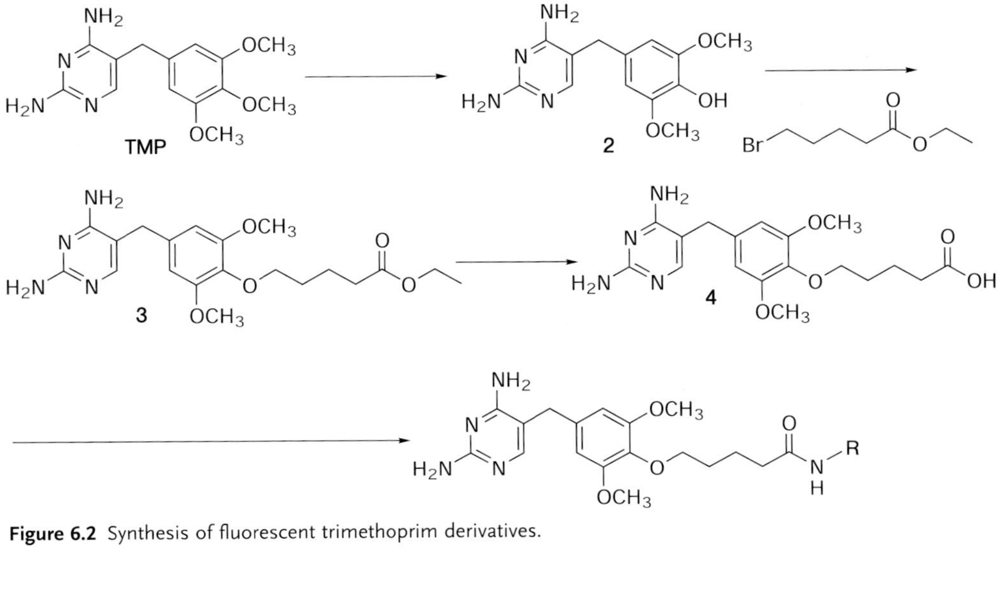

Figure 6.2 Synthesis of fluorescent trimethoprim derivatives.

A number of fluorescent TMP analogs for labeling proteins in living cells have been prepared and tested (Figure 6.1B). In general, molecules of the type TMP-linker-fluorophore can be prepared that will bind with high affinity to eDHFR. The linker moiety should be sufficiently long (about 10 atoms or more) so as not to disrupt binding to eDHFR. Linkers that comprise two or more ethylene glycol units are optimal because they increase the solubility of the molecule. The above-described reaction scheme can be adapted to prepare a 4′-alkylamino-substituted TMP derivative that can in turn be coupled to a wide variety of commercially available, amine-reactive fluorophores or affinity labels using standard peptide coupling chemistries. For routine, intracellular fluorescent labeling of eDHFR fusion proteins, red and green fluorescein-based TMP derivatives are available for purchase from Active Motif, Inc. (Carlsbad, CA, 760 431-1263, www.activemotif.com) under the LigandLink trademark.

6.4
Cell Growth and Transfection

Researchers may use their preferred method for transiently transfecting mammalian cells with plasmid DNA for overexpression of eDHFR fusion proteins. Here, Chinese Hamster Ovary (CHO-K1) cells or mouse embryonic fibroblast (MEF) cells were seeded at 10^5 cells per well into a 6-well plate. Cells were grown in Dulbeccos Modified Eagle Medium (DMEM) with 10% Fetal Bovine Serum, 15 mM HEPES, 100 IU mL^{-1} Penicillin, 100 µg mL^{-1} Streptomycin, 2 mM L-Glutamine at 37 °C and 5% CO_2. After about 18 h, adherent cells (about 80% confluent) were transfected with 2 µg of the desired plasmid DNA using Fugene 6 transfection reagent according to the manufacturers instructions. About 24 h after transfection, cells were trypsinized and reseeded onto 22 mm^2 coverslips in 6-well plates (50 000 cells/well).

6.5
Protein Labeling and Microscopy

Adherent, transfected cells were incubated in growth medium containing varying concentrations (0.2–2 µM) of fluorescent TMP derivatives (e.g., TMP-BODIPY Texas Red (TMP-BTR), LigandLink hexachlorofluorescein (TMP-HEX), or LigandLink fluorescein (TMP-FLN)) for anywhere from 10 min to 2 h depending on cell type. The incubated cells were then washed twice with PBS and mounted for imaging in growth medium without small molecule label. The stained cells were imaged using an Olympus IX81/FV-500 scanning confocal microscope with 488 nm or 543 nm laser excitation, 60X PlanFluor oil immersion objective (1.4 N.A.), and filters/dichroics appropriate for imaging fluorescein ($\lambda_{em} = 514$ nm), hexachlorofluorescein ($\lambda_{em} = 560$ nm), or Texas Red ($\lambda_{em} = 588$ nm).

6.6
Results and Discussion

Because the TMP/eDHFR ligand–receptor interaction is orthogonal to mammalian cells, it should be possible to use TMP-linked fluorophores to label eDHFR fusion proteins in any wild-type mammalian cell line. However, we have found that the concentration of the dye as well as the incubation time for staining must be optimized for a particular cell type. For example, when using CHO-K1 cells, micromolar concentrations and about 1 h incubation times are required to give maximal fluores-

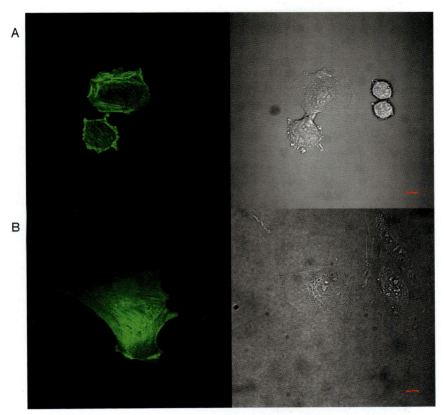

Figure 6.3 Selective chemical labeling of myosin light chain kinase fused to *E. coli* dihydrofolate reductase (MLCK-eDHFR) in wild-type mammalian cells with LigandLink fluorescein. Confocal micrographs: left column, excitation, 488 nm; right column, differential image contrast (DIC). Size bar in DIC images corresponds to 10 μm. (A) Chinese Hamster Ovary (CHO-K1) cells transiently transfected with DNA encoding MLCK-eDHFR. The cells were incubated with 2 μM LigandLink fluorescein dye for 1 h, washed twice with PBS, and mounted for imaging in growth medium without small molecule label; (B) MEF cells expressing MLCK-eDHFR. The cells were incubated with 0.2 μM LigandLink fluorescein dye for 10 min, washed twice with PBS, and mounted for imaging in growth medium without small molecule label.

cent signal (Figure 6.3A). However, we have found that with MEF cells, incubation times of 10 min with only 200 nM dye gave the best results (Figure 6.3B).

The quality of the microscopic images also depends on the type of fluorophore that is attached to the TMP ligand. The LigandLink fluorescein and hexachlorofluorescein-based dyes rapidly stain eDHFR fusion proteins with little or no detectable background

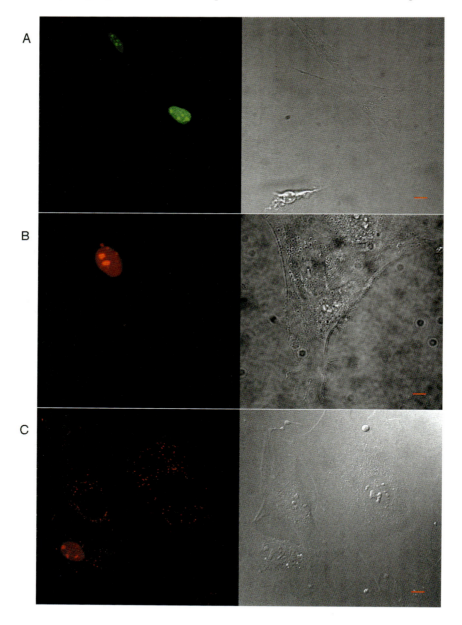

fluorescence (Figure 6.4A and B). However, when using the TMP-BODIPY Texas Red derivative described in Miller *et al.* [19] a high degree of non-specific, punctate staining is observed when cells are incubated with relatively low concentrations (about 50 nM) for short periods (<30 min, Figure 6.4C). This is likely due to the lipophilic character of the BODIPY fluorophore, which may partition to lipid bodies inside the cell (Molecular Probes, personal communication). Therefore, if researchers wish to design their own fluorescent TMP labels, they should consider the potential effects that the fluorophore's molecular structure will have on the pharmacokinetic behavior of the fluorescent TMP derivative.

Finally, because the TMP/eDHFR interaction is non-covalent, the TMP derivatives will dissociate from the eDHFR fusion proteins over time and diffuse out of the cells. This precludes experiments that require observation of fluorescently labeled eDHFR fusions over a time course of many hours or days. Nevertheless, we have detected selective fluorescent labeling of eDHFR fusion proteins in living cells for up to 2 h after the dye has been removed from the cell growth medium.

6.7
Conclusion

TMP derivatives can be used to selectively and non-covalently label eDHFR fusion proteins in living, wild-type mammalian cells. For routine fluorescent imaging applications, the cell permeable LigandLink fluorescein and hexachlorofluorescein labels can be used to tag proteins located in the plasma membrane, internal membranes or organelles, or the cytosol. The LigandLink hexachlorofluorescein label ($\lambda_{em} = 560$ nm) can be spectrally resolved from cyan or green fluorescent proteins, allowing simultaneous multi-color tagging and visualization of different target proteins in the same cell. Furthermore, the excitation spectrum of hexachlorofluorescein overlaps with the emission spectrum of GFP, suggesting its possible use as a FRET acceptor. The synthetic pathway described herein can be used to derivatize TMP with a variety of functional small molecules, including affinity tags, radical oxygen species (ROS) photosensitizers and photocrosslinking moieties, allowing the possibility of more intricate studies of protein function within living cells.

Figure 6.4 Selective chemical labeling of nucleus-targeted *E. coli* dihydrofolate reductase (eDHFR) in wild-type Mouse embryonic fibroblast (MEF) cells. MEF cells were transiently transfected with DNA encoding nucleus-targeted eDHFR (pLM1302). Confocal micrographs: left column, (A) excitation, 488 nm; (B,C), excitation 543 nm; right column, differential image contrast (DIC). Size bar in DIC images corresponds to 10 µm. (A) Adherent cells incubated with 0.2 µM LigandLink fluorescein dye for 10 min, washed twice with PBS, and mounted for imaging in growth medium without small molecule label; (B) Cells incubated with 0.2 µM LigandLink hexachlorofluorescein dye for 10 min, washed twice with PBS, and mounted for imaging in growth medium without small molecule label; (C) MEF cells incubated with 0.05 µM TMP-BODIPY Texas Red for 15 min, washed twice with PBS, and mounted for imaging in growth medium without small molecule label. Hydrophobic BODIPY Texas Red moiety partitions to lipid vesicles, resulting in punctate staining pattern.

Acknowledgements

This research was supported by the National Institutes of Health (GM071754-01).

References

1 Weber, G. (1952) Polarization of the fluorescence of macromolecules 2. Fluorescent conjugates of ovalbumin and bovine serum albumin. *The Biochemical Journal*, **51**, 155.
2 Hermanson, G.T. (1996) *Bioconjugate Techniques*, Academic Press, San Diego.
3 Tsien, R.Y. (1998) The green fluorescent protein. *Annual Review of Biochemistry*, **67**, 509.
4 Zhang, J., Campbell, R.E., Ting, A.Y. and Tsien, R.Y. (2002) Creating new fluorescent probes for cell biology. *Nature Reviews. Molecular Cell Biology*, **3**, 906.
5 Bevis, B.J. and Glick, B.S. (2002) Rapidly maturing variants of the Discosoma red fluorescent protein (DsRed). *Nature Biotechnology*, **20**, 83.
6 Campbell, R.E., Tour, O., Palmer, A.E., Steinbach, P.A., Baird, G.S., Zacharias, D.A. and Tsien, R.Y. (2002) A monomeric red fluorescent protein. *Proceedings of the National Academy of Sciences of the United States of America*, **99**, 7877.
7 Matz, M.V., Fradkov, A.F., Labas, Y.A., Savitsky, A.P., Zaraisky, A.G., Markelov, M.L. and Lukyanov, S.A. (1999) Fluorescent proteins from nonbioluminescent Anthozoa species. *Nature Biotechnology*, **17**, 969.
8 Griffin, B.A., Adams, S.R. and Tsien, R.Y. (1998) Specific covalent labeling of recombinant protein molecules inside live cells. *Science*, **281**, 269.
9 Chen, I. and Ting, A.Y. (2005) Site-specific labeling of proteins with small molecules in live cells. *Current Opinion in Biotechnology*, **16**, 35.
10 Johnsson, N. and Johnsson, K. (2003) A fusion of disciplines: chemical approaches to exploit fusion proteins for functional genomics. *Chembiochem*, **4**, 803.
11 Miller, L.W. and Cornish, V.W. (2005) Selective chemical labeling of proteins in living cells. *Current Opinion in Chemical Biology*, **9**, 56.
12 Keppler, A., Pick, H., Arrivoli, C., Vogel, H. and Johnsson, K. (2004) Labeling of fusion proteins with synthetic fluorophores in live cells. *Proceedings of the National Academy of Sciences of the, United States of America*, **101**, 9955.
13 Miller, L.W., Sable, J., Goelet, P., Sheetz, M.P. and Cornish, V.W. (2004) Methotrexate conjugates: a molecular in vivo protein tag. *Angewandte Chemie (International Edition in English)*, **43**, 1672.
14 Marks, K.M., Braun, P.D. and Nolan, G.P. (2004) A general approach for chemical labeling and rapid, spatially controlled protein inactivation. *Proceedings of the National Academy of Sciences of the, United States of America*, **101**, 9982.
15 George, N., Pick, H., Vogel, H., Johnsson, N. and Johnsson, K. (2004) Labeling of cell surface proteins with chemically diverse compounds. *Journal of the American Chemical Society*, **126**, 8896.
16 Yin, J., Liu, F., Li, X. and Walsh, C.T. (2004) Labeling proteins with small molecules by site-specific posttranslational modification. *Journal of the American Chemical Society*, **126**, 7754.
17 Chen, I., Howarth, M., Lin, W. and Ting, A.Y. (2005) Site-specific labeling of cell surface proteins with biophysical probes using biotin ligase. *Nature Methods*, **2**, 99.
18 Howarth, M., Takao, K., Hayashi, Y. and Ting, A.Y. (2005) Targeting quantum dots to surface proteins in living cells. *Proceedings of the National Academy of*

Sciences of the, United States of America, **102**, 7583.

19 Miller, L.W., Cai, Y., Sheetz, M.P. and Cornish, V.W. (2005) In vivo protein labeling with trimethoprim conjugates: a flexible chemical tag. *Nature Methods*, **2**, 255.

20 Sasso, S.P., Gilli, R.M., Sari, J.C., Rimet, O.S. and Briand, C.M. (1994) Thermodynamic study of dihydrofolate reductase inhibitor selectivity. *Biochimica et Biophysica Acta*, **1207**, 74.

21 Calloway, N., Choob, M., Sanz, A., Miller, L.W. and Cornish, V.W. (2007) Optimized fluorescent trimethoprim derivatives for in vivo protein labeling. *Chembiochem*, **8**, 767.

22 Ausubel, F., Brent, R., Kingston, R., Moore, D., Seidman, J., Smith, J. and Struhl, K. (1995) *Current Protocols in Molecular Biology*, John Wiley & Sons, Inc., New York.

23 Poperechnaya, A., Varlamova, O., Lin, P.J., Stull, J.T. and Bresnick, A.R. (2000) Localization and activity of myosin light chain kinase isoforms during the cell cycle. *The Journal of Cell Biology*, **151**, 697.

7
Phosphopantetheinyl Transferase Catalyzed Protein Labeling and Molecular Imaging

Norman J. Marshall and Jun Yin

7.1
Introduction

Protein labeling is an essential tool for studying the biological functions of proteins *in vitro* and in living cells or organisms. A variety of target-specific labeling technologies have been developed that make use of the unique properties of naturally existing proteins, including autofluorescence or affinity for functional small molecules or other proteins. Generally, these strategies require the construction of recombinant protein fusions in order to establish the physical linkage between the target protein and the protein or peptide tag serving as an affinity epitope or fluorescence reporter. Various protein and peptide affinity tags, including maltose binding protein (MBP) [1], glutathione S-transferase (GST) [2], 6× Histidine tag [3], c-myc tag [4, 5] and FLAG tag [6], have been widely used to facilitate the purification and labeling of the fusion protein based on their binding affinity with small molecule ligands or antibodies [7]. Fluorescence reporters, such as the green fluorescent protein (GFP) of *Aequorea victoria*, may also be appended to a protein of interest, allowing real-time imaging of the distribution, translocation and degradation of the protein fusions inside the living cell [8–10].

Recently, a number of protein posttranslational modification enzymes have been used for site-specific protein labeling by catalytically conjugating small molecule probes to peptide tags fused to the target proteins. Protein posttranslational modification (PTM) is a universal mechanism for the delicate control of a protein's structure and function, and plays crucial roles in virtually all aspect of cellular life – from signal transduction to cell division and apoptosis. Often PTM diversifies the structure of the proteins by installing new chemical entities that are not provided by the 20 proteinogenic amino acids in the peptide chain. This facilitates catalysis, binding and other functions of the modified protein [11]. For example, phosphorylation [12] and glycosylation [13] induce conformational changes in target proteins, modulating their enzymatic activities and interactions with other molecules. Protein biotinylation [14], lipoylation [15], and phosphopantetheinylation [16] install crucial functional groups which directly participate in substrate turnover at the active site of biosynthetic

Probes and Tags to Study Biomolecular Function. Lawrence W. Miller (Ed.)
Copyright © 2008 WILEY-VCH Verlag GmbH & Co. KGaA, Weinheim
ISBN: 978-3-527-31566-6

enzymes. Protein lipidation designates the subcellular localization of the modified target and its trafficking pathway within the cell [17]. Since protein labeling is essentially a posttranslational modification event, various PTM enzymes might offer a unique opportunity to be used as tools for the attachment of diverse small molecule probes to the target protein and facilitate the elucidation of its function in the complex proteome.

To label target proteins by posttranslational modification, protein or peptide substrates of the PTM enzymes need to be fused to the target protein, followed by the enzymatic attachment of chemical probes such as fluorophores to the substrate tag. So far, a number of methods in this category have emerged, including biotin ligase [18–20], human O^6-alkylguanine-DNA alkyltransferase [21], transglutaminase [22, 23], sortase [24], cutinase [25] and Sfp or AcpS phosphopantetheinyl transferases (PPTases) [26–28], for the covalent attachment of small molecule probes to the target proteins. Compared to labeling proteins by expressing fusions with fluorescent proteins such as GFP, these methods add a level of control over the type of label appended to the target, as well as the stage at which the label is to be introduced. For example, the interaction between biotin and streptavidin has been widely used for protein pull-down and isolation experiments. To take advantage of this high affinity interaction and label proteins with biotin, a short peptide substrate of the E. coli biotin ligase, BirA, was identified for the biotinylation of the peptide at a specific Lys residue [19, 20]. The short peptide substrate of BirA, known as AP peptide, is 14 residues in length – much smaller in size than the native substrate of BirA, biotin carboxyl carrier protein (BCCP) [19]. In the presence of ATP, BirA catalyzes the formation of the biotin-AMP intermediate, the breakdown of which leads to the biotinylation of the AP peptide that is fused to the target protein. Subsequently, the biotin affinity handle may be used to isolate the target protein, or attach a fluorescence probe to the target protein using a fluorophore-conjugated streptavidin derivative [18–20]. Recently, phage display has been used to identify a peptide tag specifically biotinylated by yeast biotin ligase. In combination with the AP peptide tag, different cell surface proteins may be labeled with quantum dot-conjugated streptavidin for imaging purposes [29].

In another example, human O^6-alkylguanine-DNA-transferase (hAGT) is used for site specific protein labeling. hAGT is a DNA repair protein of 207 residues in size [30] that irreversibly transfers O^6 substituents from alkylguanine derivatives to a conserved Cys residue of its own. Since the self-modification of hAGT by alkyl transfer is permanent, hAGT has been used as a protein tag to be fused to the target protein. Labeling is then achieved by transferring biotin and fluorophore functionalities to the hAGT tag from the corresponding O^6-benzylguanine derivatives [21, 31, 32]. This system has been demonstrated in AGT deficient CHO cells for imaging a variety of proteins fused to hAGT, including β-galacatosidase, α-tubulin, and simian virus 40 large T antigen nuclear localization sequences [21, 31, 32]. In each case, an AGT deficient cell line is needed to reduce background labeling of endogenous AGT [21, 31, 32].

Transglutaminases have also been used for protein labeling, by enzymatically linking glutamine to a free amine that is attached to a small molecule probe [22, 23, 33, 34]. These enzymes are found in prokaryotes and eukaryotes alike, and their native activity is to carry out a variety of protein cross-linking reactions between Glu and Lys residues on different proteins [35]. Usually, transglutaminases are specific to their glutamine substrates, yet tolerant of various amine substrates [35] allowing their application for protein labeling. Recently, transglutaminases from both microbial sources [22, 33, 34] and guinea pig liver [23] have been used to conjugate target proteins with small molecule fluorophores or fluorescent protein tags.

There are several common considerations for a successful protein labeling method. Firstly, the protein tag must not interfere with the folding or native function of the protein of interest. Therefore, the protein tag is, ideally, small with respect to the target protein. Secondly, the labeling technique should be highly specific toward the peptide tag fused to the target protein in order to reduce background signal. Finally, the method should be applicable to a variety of small molecule labels with diverse chemical entities so that the structure and function of the target protein can be probed from different perspectives. This chapter focuses on an efficient method for live cell, site-specific protein labeling based on the posttranslational modification reaction catalyzed by phosphopantetheinyl transferases – including Sfp of *Bacillus subtilis* origin [36], and AcpS of *Escherichia coli* origin [37, 38]. These enzymes transfer a phosphopantetheinyl linker from coenzyme A to the conserved Ser residue of peptidyl or acyl carrier proteins (PCP or ACP, respectively). The carrier proteins may be used as protein tags to be fused to target proteins and subsequently labeled with a fluorophore-coenzyme A (CoA) conjugate [26, 27]. Short peptide tags have been developed showing orthogonal substrate specificity for Sfp and AcpS modification, allowing differential labeling of distinct target proteins on the surface of the same cell [39, 40]. This system benefits from the relatively small size of the peptide tags, fast reaction kinetics and the diversity of the small molecule probes that can be covalently transferred to the carrier protein or peptide tags from their CoA conjugates [26].

7.2
Protein Posttranslational Modification by Phosphopantetheinyl Transferases

As their name suggests, the native function of PPTases is the posttranslational phosphopantetheinylation of the carrier protein modules in nonribosomal peptide synthetases (NRPS), polyketide synthases (PKS), and fatty acid synthases (FAS) [41]. NRPS and PKS are large, modular enzyme assemblies responsible for the production of an enormous variety of nonribosomal peptide and polyketide natural products [41–43] (Figure 7.1). In order for the enzymes to be active, a 20 Å, 4′-phosphopantetheinyl (Ppant) linker must be installed on carrier protein domains embedded within the NRPS or PKS clusters by PPTases [36, 41, 43]. The Ppant linker is added to a specific serine residue in the carrier proteins, and functions as a "swinging arm" to provide an anchor point for the growing peptide or polyketide chain as it is elongated along the NRPS or PKS assembly line [41] (Figure 7.2).

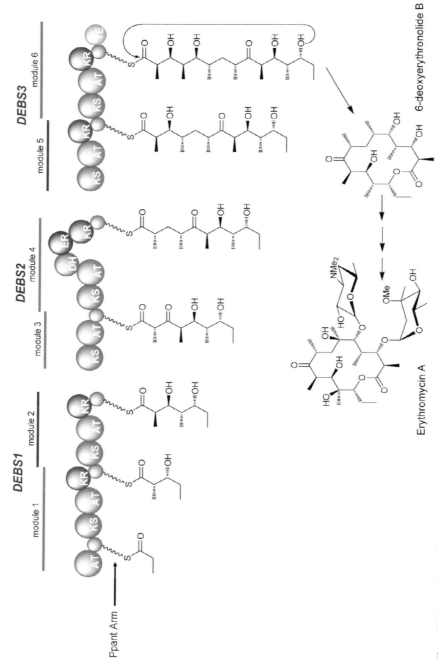

Figure 7.1 Biosynthesis of Erythromycin A precursor 6-deoxyerythronolide B via polyketide synthase DEBS [42]. The enzyme assembly is arranged into modules, with the growing polyketide anchored to carrier proteins by the 20 Å Ppant arm (as shown by wavy linkers).

7.2 Protein Posttranslational Modification by Phosphopantetheinyl Transferases | 125

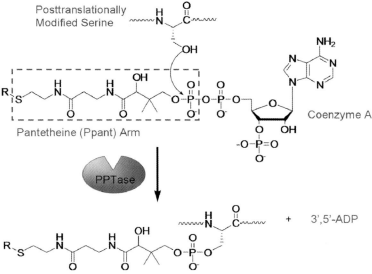

Figure 7.2 Target protein fusion labeling reaction catalyzed by phosphopantetheinyl transferases. The 20 Å Ppant arm is transferred to the conserved serine residue of the carrier protein or peptide tag fused to the protein of interest from CoA derivatives by Sfp or AcpS PPTases.

PPTases are classified into two general categories: Sfp-type, which modify carrier proteins embedded in NRPS, and AcpS-type, which modify carrier proteins embedded in PKS and FAS [36]. Sfp-type PPTases function as monomers of about 230 residues in size, with twofold pseudosymmetry. The crystal structure of Sfp in complex with CoA, has been solved, and it shows each "half" of the Sfp enzyme consists of three anti-parallel β-sheets and one long α-helix, separated by a long loop region [44] (Figure 7.3). The β-sheets of each "half" are positioned together at the core of the enzyme, with the long helices positioned adjacent. The loop regions lie to the periphery and contain short helices. The active site is formed between the two "halves." CoA is positioned in the active site such that the 3′ phosphate and the 5′ α and β phosphates interact with several basic amino acids, as well as a magnesium ion. The Ppant arm is bent away from the surface of the enzyme and protrudes out into the solvent [44] (Figure 7.3). This probably accounts for the promiscuity of Sfp toward CoA substrates containing many different chemical entities conjugated to the free thiol of the Ppant arm [45–48], since they do not contact the enzyme. Sfp-type PPTases have not yet been co-crystallized with their PCP substrates, so the exact binding mode between Sfp and the carrier proteins is unknown.

AcpS-type PPTases function as homotrimers of about 120 residues each, forming an active catalytic site between each of the monomeric subunits, as seen in the crystal structures [49, 50] (Figure 7.4). Like the Sfp-type PPTases, each subunit consists of three anti-parallel β-sheets and a long α-helix, separated by smaller helices and a long loop region. In this case, the β-sheets come together at the core of the homotrimer,

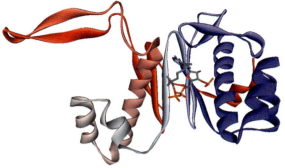

Figure 7.3 *Bacillus subtilis* Sfp co-crystalized with coenzyme A [44]. Sfp possesses twofold pseudosymmetry, with the CoA binding pocket in the middle (shown as stick model). The Ppant arm of CoA is bent away from the surface of the enzyme and protrudes into the solvent, which is probably responsible for the high promiscuity toward small molecules appended to the CoA terminal thiol.

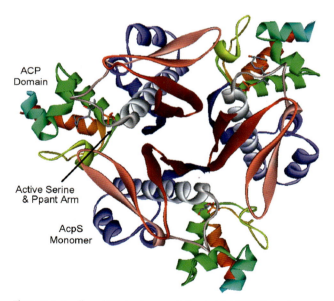

Figure 7.4 *Bacillus subtilis* AcpS co-crystalized with ACP [50]. AcpS exists as a homotrimer, with the ACP binding sites in the cleft between each monomer. The AcpS/ACP structure shows that helix 2 is largely responsible for enzyme recognition. The active site serine residue lies at the top of helix 2, and is shown covalently modified with Ppant group.

which effectively hides key hydrophobic residues. The long α-helix is adjacent to the β-sheets, and it is surrounded by the other helices and the loop region. Also analogous to Sfp-type PPTases, the catalytic site lies in the cleft between the α-helices [49, 50]. However, only two of the active sites appear to bind the CoA ligand [49] (Figure 7.4). The CoA is again held in place by several basic residues and the α and β phosphates bind a magnesium ion. The Ppant linker rests on the enzyme surface and is not bent out into the solvent [50] (Figure 7.4). Though the monomeric subunits bear striking structural similarity to the Sfp-type PPTases, there is very little sequence homology [49, 50]. The AcpS of *Bacillus subtilis* has been co-crystallized with its ACP substrate to show that the long α-helix of AcpS is predominantly responsible for binding the carrier protein [50] (Figure 7.4).

PPTases such as Sfp and AcpS have shown substantial substrate tolerance toward entities covalently attached to CoA through the terminal thiol. It has been demonstrated that fluorophores, biotin, sugars, peptides and porphyrin are all compatible with the PPTase transfer mechanism, and as such, may be used as labels for carrier protein fusions [45–48, 51–53]. Toward their carrier proteins, however, Sfp and AcpS exhibit different specificities. While AcpS is highly specific toward the ACP involved in PKS and FAS, Sfp catalyzes the Ppant transfer to both the PCP embedded in NRPS, and ACP [54, 55]. This selectivity may be due to the difference in surface charge between the carrier proteins: ACP, with a pI of 3.8, is more negatively charged than PCP, which has a pI of 7.6 [54, 56]. AcpS and Sfp both compliment their carrier proteins, having pI values of 9.6 and 5.6, respectively [54].

Carrier proteins PCP and ACP are the native substrates of PPTases. They are relatively small, autonomously folding proteins of about 80 to 100 residues to which the Ppant prosthetic arm is attached at the active Ser residue [43]. Both PCP and ACP are globular proteins composed of three or four α-helices separated by flexible loop regions [56, 57]. In the carrier proteins, the Ppant modified Ser residue resides in a short loop region N-terminal to helix 2, which has been suggested to play an important role for PPTase recognition [54] (Figure 7.5). NMR study has shown that PCP changes conformation between its apo and holo form [58]. The apo form, designated "A," is less organized, with the loop region between helices 1 and 2 extending into the helix 2 residues. Upon Ppant transfer, however, a slow transition between the "A" form and the holo, "H" form takes place (Figure 7.6). The protein becomes more structured, with more well-defined helices. Mutating the active Ser residue to Ala locks PCP into the "A" state, which suggests that it is indeed Ppant modification that causes the conformation change [58]. NMR study of holo ACP has shown a similar transition between two conformations, however, the exact cause is unknown [59].

7.3
Protein Labeling Via Carrier Protein Fusions

The full-length carrier proteins have been employed as protein tags for site-specific labeling by the PPTase enzymes. The reaction is very efficient, with sufficient labeling

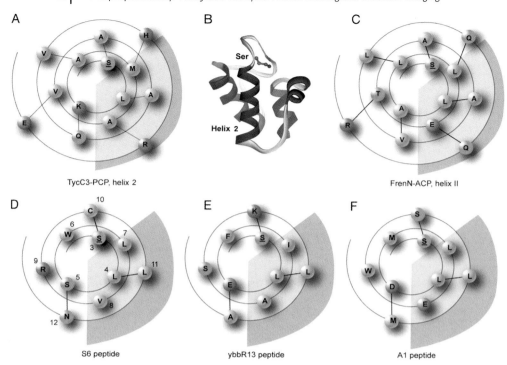

Figure 7.5 Helical wheel plots of the helix 2 residues of PCP and ACP, and the short peptide tags, ybbR, S6, and A1 [40]. The Ppant modified Ser residue is underlined for each plot, and designated residue 3 in plot (D). (A) Full-length TycC3-PCP helix 2 [56] (B) NMR structure of TycC3 PCP indicating helix 2 and the active Ser residue (C) Full-length FrenN-ACP helix 2 [59] (D) Short peptide tag S6 for Sfp catalyzed phosphopantetheinylation [40] (E) Short ybbR tag from *B. subtilis* genome identified as efficient substrate of Sfp [39] (F) Short peptide tag A1 for AcpS catalyzed phosphopantetheinylation [40]. Comparison of the shaded, solvent-exposed surface shows strong agreement between the native carrier proteins and their respective peptide tag substitutes. As labeled in (D), the residues 4, 7, 8, and 11 are vital for PPTase recognition and binding.

achieved after 15 min at submicromolar concentrations of both the PPTase enzyme and fluorophore-conjugated CoA [26]. In one example, PCP was fused to the extracellular C-terminus of transferrin receptor 1 (TfR1), and the TfR1-PCP fusion at the cell surface was labeled with Alexa 488 by Sfp catalyzed covalent transfer of the small molecule fluorophore from its CoA conjugate to the PCP tag [53] (Figure 7.7). Upon binding of the Alexa 568 labeled transferrin ligand to the cell surface TfR1-PCP fusion, fluorescence resonance energy transfer (FRET) was demonstrated between the donor fluorophore Alexa 488 on TfR1 and the acceptor fluorophore Alexa 568 on the transferrin ligand (Figure 7.8). Additional iron uptake and transferrin binding assays showed that PCP did not interfere with the normal function of TfR1 [53]. Furthermore, real-time observation of endocytosis of the transferrin receptor was

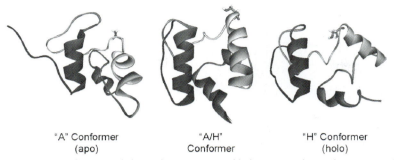

Figure 7.6 Conformational change between apo and holo PCP, as observed in NMR studies [58].

possible, showing the co-migration of fluorescently labeled TfR1 and transferrin from the cell surface to the early endosome [53] (Figure 7.8).

ACP has also been used as a protein fusion to tag a cell surface receptor. In one study, ACP was appended to the N-terminus of human Nerokinin-1 receptors, NK1R, and labeled by AcpS with a variety of fluorophores conjugated to CoA [51, 60]. By applying a labeling mixture of Cy3-CoA and Cy5-CoA, AcpS could be used to label the ACP-NK1R fusions simultaneously with two fluorophores [60]. A homodimer between NK1R receptors thus produced a FRET signal between the Cy3 donor fluorophore and the Cy5 acceptor fluorophore when NK1R was overexpressed. The absence of FRET under endogenous NK1R expression levels was used to show that the NK1 receptors are naturally monomeric [60]. This labeling technique can also be

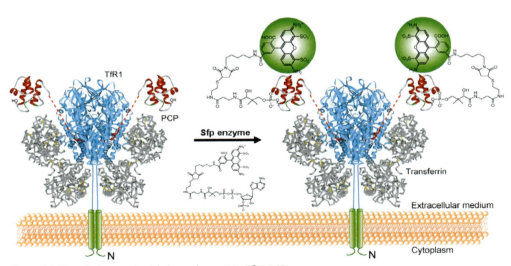

Figure 7.7 Transferrin receptor labeling scheme [53]. TfR1-PCP fusion is expressed on the surface of TRVb cells and treated with an Alexa 488-CoA conjugate and Sfp. The fluorescent labeled TfR1 may then be visualized in the live cell.

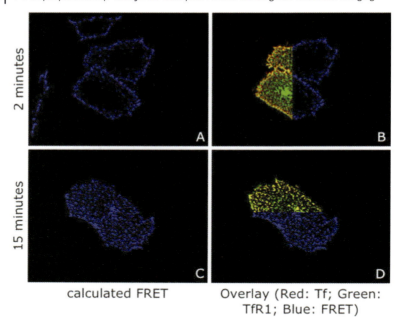

Figure 7.8 Calculated FRET interactions between TfR1-PCP labeled with an Alexa 488-CoA conjugate and the transferrin ligand labeled with Alexa 568 (Tf-Alexa 568) in TRVb cells [26, 53]. The measurements were performed 2 min and 15 min after addition of Tf-Alexa 568.
(A) Calculated relative intensity of FRET between donor and acceptor fluorophores after 2 min (B) Overlay of TfR1-PCP (green), Tf–Alexa 568 (red) and calculated FRET (blue) after 2 min (C) Calculated FRET after 15 min (D) FRET and confocal overlay 15 min after the addition of Tf–Alexa 568.

applied to time-resolved visualization of multiple generations of the ACP-target protein fusions.

For example, ACP was appended to the *S. cerivisiae* cell wall protein, Sag1p, and catalytically labeled by AcpS with a Cy3-CoA conjugate *in vivo* [52]. Time was given for the yeast cells to bud, and then the ACP-Sag1p fusions were labeled with Cy5-CoA conjugates. The physical distribution of the different fluorophores clearly identified areas of reproductive growth [52].

Besides its application for imaging purposes, PPTases have also been used to immobilize various proteins on a microarray [47] and to attach small molecule libraries to the surface of phage for DNA encoding [48]. To construct a protein microarray, PCP-protein fusions were expressed in *E. coli* and specifically labeled in cell lysates by Sfp catalyzed biotin transfer from its CoA conjugates. The biotinylated target protein fusions were then spotted onto an avidin glass slide to print the protein microarray. After spotting, the *E. coli* proteins in the cell lysates were washed away, and the immobilized PCP fusions were visualized by treating the slide with target-specific antibodies conjugated to different small molecule fluorophores, respectively [47].

To attach small molecules to the surface of phage, the PCP domain was displayed as a fusion to M13 phage capsid protein pIII. Sfp was then used to append the small molecules to the phage displayed PCP domain by covalent transfer from their CoA conjugates. In this way, a DNA barcode encapsulated within the phage particle is physically linked to the small molecule it encodes. After affinity selection with a target receptor, the DNA bar code in the selected phage particle may be amplified and decoded by a DNA array to determine the identity of the small molecule selected from the library [48].

7.4
Orthogonal Protein Labeling by Short Peptide Tags

Despite the relatively small size of the PCP and ACP fusions (80–100 residues) compared to hAGT (207 residues) [21] or the Green Fluorescent Protein (238 residues) [8], they are still, unfortunately, large with respect to many peptide tags routinely used as affinity epitopes. Therefore, to decrease the size of the target protein fusions, an 11 residue peptide, ybbR, was recently identified from a genomic library of *Bacillus subtilis* by phage display as an efficient substrate of Sfp [39]. During phage selection, a genomic library of *B. subtilis* was displayed on the phage surface, followed by the addition of Sfp and biotin-CoA for the site-specific labeling of the phage displayed Sfp substrate proteins with biotin. Subsequently, biotinylated phage particles are enriched by binding to immobilized streptavidin. From this selection, a truncated ybbR protein encoded in the *B. subtilis* genome was identified as a substrate for Sfp modification. The minimum sequence required for modification was found to be the 11 residue sequence of ybbR (DS̲LEFIASKLA). In addition to the underlined Ser as the site of Ppant attachment, the ybbR peptide contains the flanking Asp and Leu residues, corresponding to the semi-conserved DSL [61] motif from the site of Ppant modification in the native carrier proteins (Figure 7.5). Furthermore, it has been demonstrated that the ybbR tag may be fused to either the N- or C-terminus of the target protein – and it can even be used as an internal tag, at an accessible position within the protein [39].

In a further study to identify short peptide substrates specific for Sfp or AcpS modification, peptide libraries displayed on the surface of M13 phage were selected in parallel by Sfp and AcpS catalyzed biotin-CoA labeling [40]. The peptide libraries with the sequence GDS(L/I)XXXXXXXX, where X = any of the 20 protein residues, were constructed by retaining the conserved DS(L/I) motif found at the site of Ppant modification in the native carrier proteins and randomizing the subsequent eight residues corresponding to helix 2 in the full length carrier protein [61] (Figure 7.5). The final size of the peptide library displayed on the phage surface was approximately 1×10^9, 50-fold smaller than the theoretical diversity of the library (5×10^{10}) calculated from the number of randomized residues in the peptide. The randomized library was inserted into phagemid pComb3H [62] at the N-terminus of M13 phage capsid pIII protein. Peptides were selected by Sfp and AcpS separately from the randomized library. After five rounds of phage selections, two 12 residue peptides – S6 and A1 – were enriched as orthogonal substrates of Sfp and AcpS from the peptide library. The catalytic specificity (k_{cat}/K_m) of Sfp-catalyzed S6 labeling is more than 440-fold higher than AcpS-catalyzed S6 labeling. Conversely, the specificity (k_{cat}/K_m) of AcpS-catalyzed A1 labeling is more than 30-fold higher than Sfp-catalyzed A1 labeling [40], suggesting the orthogonality of S6 and A1 labeling catalyzed by Sfp and AcpS, respectively.

In order to compare the topology of the selected peptides to helix 2 in the full length carrier proteins, the sequences of S6, A1 and ybbR were drawn in helical wheel plots and compared with the plots for the helix 2 of TycC3-PCP [56] and FrenN-ACP [59] (Figure 7.5). The comparison identified significant homology between the peptide residues and carrier protein residues populating the solvent exposed surface of helix 2 (Figure 7.5, residues 4, 7, 8 and 11 shown in gray). For example, Leu and Val residues occupy positions 4, 7, 8, and 11 of peptide S6, forming one side of the helix, quite similar to the helix 2 in TycC3-PCP with small hydrophobic residues Leu, Ala and Met at the same positions. The A1 peptide contains a negatively charged Glu residue at position 8 on the side of helix with three Leu residues at positions 4, 7, and 11. This corresponds to helix 2 of FrenN-ACP, with Glu residue at position 8 and Leu and Ala residues at the other positions. From the striking similarities between the selected peptide sequences and the PCP or ACP sequences, it is clear that solvent-exposed residues at 4, 7, 8, and 11 play important roles interacting with the Sfp or AcpS enzyme active site.

We have demonstrated that S6tag/Sfp and A1tag/AcpS can be used as orthogonal pairs for site-specific labeling of differentially tagged target proteins on live cell surfaces [40]. Additionally, they may be appended to either the N- or C-terminus of the target protein. The A1 peptide was fused to the extracellular C-terminus of TfR1 and expressed in TRVb cells. AcpS was then used to catalytically transfer a fluorophore, Alexa 488, from its CoA conjugate to the A1 tag. After treating the cells with fluorophore Alexa 568 conjugated transferrin, the two fluorophores were observed to be co-localized on the surface of the cells [40]. Thus, the A1 tag is a suitable substitute for full-length carrier protein in PPTase catalyzed protein labeling (*vide supra*). Similarly, the S6 tag was fused to the extracellular N-terminus of epidermal growth factor receptor (EGFR) and expressed in HeLa cells. A Texas red-CoA conjugate was then catalytically transferred to the S6 tag by Sfp. After treating the cells with fluorophore Alexa 488 conjugated epidermal growth factor (EGF), the colocalization between the EGF ligand and receptor observed by confocal microscopy showed that the S6 tag and its Ppant linker did not interfere with the normal function of EGFR [40].

In another experiment, the N-terminus S6-EGFR fusion and the C-terminus TfR1-A1 fusion were coexpressed in HeLa cells [40]. The cells were first treated with AcpS and the Alexa 488-CoA conjugate to label the TfR1-A1 fusion. Then, the cells were treated with Sfp and the Texas red-CoA conjugate (Figure 7.9). As expected, confocal microscopy showed that two populations of cell surface receptors were sequentially labeled with two different types of fluorophores [40]. To further test the orthogonality of the S6 and A1 tags toward their mismatched PPTases, cells expressing only the Tfr1-A1 fusion were treated with Sfp and the Texas red-CoA conjugate. Conversely, cells expressing only the S6-EGFR fusion were treated with AcpS and the Alexa 488-CoA conjugate. Mismatched labeling was extremely low for both tags [40].

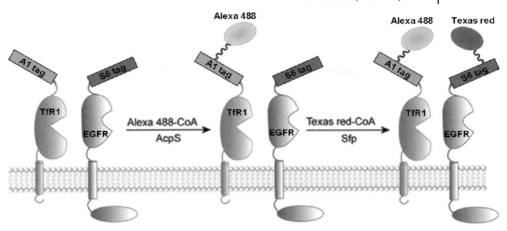

Figure 7.9 Tandem labeling of TfR1-A1 and S6-EGFR protein fusions [40]. The TfR1-A1 and S6-EGFR fusions are co-expressed on the surface of HeLa cells. The cells are first treated with an Alexa 488-CoA conjugate and AcpS to selectively label the TfR1 receptor fusions, and then treated with a Texas red-CoA conjugate and Sfp to selectively label the EGFR receptor fusions.

7.5
Summary and Perspectives

In conclusion, phosphopantetheinyl transferases offer a powerful and versatile tool for site-specific protein labeling. They possess several virtues of a good protein labeling method, including small size of the peptide tag for fusion protein construction; diversity of the small molecule probes that can be covalently conjugated to the peptide tag; high efficiency and specificity of the labeling reaction; compatibility with live cell labeling and imaging; and, the feasibility of orthogonal labeling using different PPTase/peptide tag pairs for distinct target proteins on the surface of the same cell. In this method, PPTases transfer small molecule probes to native carrier protein tags of a relatively small size (80–100 residues in length) [47, 51]. Additionally, short, 12 residue peptide tags (ybbR, S6, and A1) have been developed that can be used in place of the full-length carrier proteins for Sfp and AcpS catalyzed protein labeling [39, 40]. The PPTase enzymes also demonstrate impressive substrate promiscuity toward chemical groups of diverse structures and functionalities appended to the terminal thiol of CoA [46, 48, 63]. Thus, a variety of small molecule probes can be used to label a target protein based on the single tag modification catalyzed by PPTases. PPTases have proven to be compatible for protein labeling in different cell culture media, specifically modifying the target protein tags anchored on the cell surface [51, 53], phage particles [48], or in the cell lysates [47]. Furthermore, the development of the S6 and A1 tags for orthogonal protein labeling has opened new venues for imaging multiple target proteins and studying their interactions in the living cell.

The S6 and A1 peptide tags are modified by Sfp and AcpS with k_{cat} values around 3 min^{-1}, similar to that of the full length PCP and ACP proteins [40]. However, the K_m values for PPTase catalyzed S6 and A1 modification are around 100 μM – more than 50-fold higher than that for the full length PCP and ACP domains [40]. In practice, the lower K_m values of the peptide tags with respect to the full length carrier proteins may create difficulties for labeling cell surface receptors of lower abundance with the S6 and A1 peptide tags. Either high enzyme concentration or long reaction time would be needed; however, both would increase the background labeling of the cell. Thus, it would be useful to further evolve the peptide sequence of A1 and S6 in order to identify short peptide substrates with lower K_m values for more efficient protein labeling by AcpS and Sfp. Ultimately, a K_m for the peptide tag modification reaction comparable to that of the full length PCP or ACP proteins is desired. The tags should maintain their orthogonality while increasing sensitivity for low concentration applications.

Additionally, the impermeability of the cell membrane to CoA provides both an advantage and a limitation to this technique. For proteins expressed on the surface of a cell, labeling may be achieved with extremely low background, thanks to the exclusion of CoA conjugates from the cytosol [26]. As a limitation, the exclusion of PPTase labeling within the cell might be overcome by engineering a biosynthetic precursor of CoA that already contains the desired label molecule, as demonstrated in *E. coli*. The complete, labeled CoA substrate would then be assembled enzymatically in the cytosol [64]. This, however, requires the expression of multiple enzymes. Alternatively, a fluorophore-CoA precursor in which covalent protecting groups mask the triple negative charge of the CoA phosphates may be explored. With the phosphates in CoA neutralized as phosphoesters, the linker would be able to diffuse into the cell across the membrane and subsequently be hydrolyzed to yield the desired fluorophore-CoA [65]. Together, the labeling technique would require fusing short peptide tags to the target proteins, treating the cells with the modified coenzyme A, and then labeling with the respective PPTase enzymes. Such labeling systems will be attractive and important tools for imaging intracellular proteins and elucidating their biological functions.

References

1 Di Guan, C., Li, P., Riggs, P.D. and Inouye, H. (1988) Vectors that facilitate the expression and purification of foreign peptides in *Escherichia coli* by fusion to maltose-binding protein. *Gene*, **67**, 21–30.

2 Smith, D.B. and Johnson, K.S. (1988) Single-step purification of polypeptides expressed in *Escherichia coli* as fusions with glutathione S-transferase. *Gene*, **67**, 31–40.

3 Hochuli, E., Bannwarth, W., Dobeli, H., Gentz, R. and Stuber, D. (1988) Genetic approach to facilitate purification of recombinant proteins with a novel metal chelate adsorbent. *Nature Biotechnology*, **6**, 1321–1325.

4 Evan, G.I., Lewis, G.K., Ramsay, G. and Bishop, J.M. (1985) Isolation of monoclonal antibodies specific for human c-myc proto-oncogene product. *Molecular and Cellular Biology*, **5**, 3610–3616.

5 Munro, S. and Pelham, H.R. (1986) An Hsp70-like protein in the ER: identity with

the 78 kd glucose-regulated protein and immunoglobulin heavy chain binding protein. *Cell*, **46**, 291–300.
6. Hopp, T.P., Prickett, K.S., Price, V.L., Libby, R.T., March, C.J., Cerretti, D.P., Urdal, D.L. and Conlon, P.J. (1988) A short polypeptide marker sequence useful for recombinant protein identification and purification. *Nature Biotechnology*, **6**, 1204–1210.
7. Terpe, K. (2003) Overview of tag protein fusions: from molecular and biochemical fundamentals to commercial systems. *Applied Microbiology and Biotechnology*, **60**, 523–533.
8. Tsien, R.Y. (1998) The green fluorescent protein. *Annual Review of Biochemistry*, **67**, 509–544.
9. Kain, S.R., Adams, M., Kondepudi, A., Yang, T.T., Ward, W.W. and Kitts, P. (1995) Green fluorescent protein as a reporter of gene expression and protein localization. *BioTechniques*, **19**, 650–655.
10. Yang, T.T., Cheng, L. and Kain, S.R. (1996) Optimized codon usage and chromophore mutations provide enhanced sensitivity with the green fluorescent protein. *Nucleic Acids Research*, **24**, 4592–4593.
11. Walsh, C.T. (2006) *Posttranslational Modification of Proteins: Expanding Nature's Inventory*, 1st edn. Roberts and Company Publishers, Greenwood Village, CO, USA.
12. Johnson, L.N. and Lewis, R.J. (2001) Structural basis for control by phosphorylation. *Chemical Reviews*, **101**, 2209–2242.
13. Wormald, M.R., Petrescu, A.J., Pao, Y.L., Glithero, A., Elliott, T. and Dwek, R.A. (2002) Conformational studies of oligosaccharides and glycopeptides: complementarity of NMR, X-ray crystallography, and molecular modeling. *Chemical Reviews*, **102**, 371–386.
14. Knowles, J.R. (1989) The mechanism of biotin-dependent enzymes. *Annual Review of Biochemistry*, **58**, 195–221.
15. Gunsalus, I.C. (1953) The chemistry and function of the pyruvate oxidation factor (lipoic acid), *Journal of Cellular Physiology Supplement*, **41**, 113–136.
16. McCarthy, A.D. and Hardie, D.G. (1984) Fatty acid synthase- an example of protein evolution by gene fusion. *Trends in Biochemical Sciences*, **9**, 60–63.
17. Bhatnagar, R.S. and Gordon, J.I. (1997) Understanding covalent modifications of proteins by lipids: where cell biology and biophysics mingle. *Trends in Cell Biology*, **7**, 14–21.
18. Chen, I., Howarth, M., Lin, W. and Ting, A.Y. (2005) Site-specific labeling of cell surface proteins with biophysical probes using biotin ligase. *Nature Methods*, **2**, 99–104.
19. Beckett, D., Kovaleva, E. and Schatz, P.J. (1999) A minimal peptide substrate in biotin holoenzyme synthetase-catalyzed biotinylation. *Protein Science*, **8**, 921–929.
20. Kulman, J.D., Satake, M. and Harris, J.E. (2007) A versatile system for site-specific enzymatic biotinylation and regulated expression of proteins in cultured mammalian cells. *Protein Expression and Purification*, **52**, 320–328.
21. Keppler, A., Gendreizig, S., Gronemeyer, T., Pick, H., Vogel, H. and Johnsson, K. (2003) A general method for the covalent labeling of fusion proteins with small molecules *in vivo*. *Nature Biotechnology*, **21**, 86–89.
22. Kamiya, N., Tanaka, T., Suzuki, T., Takazawa, T., Takeda, S., Watanabe, K. and Nagamune, T. (2003) S-peptide as a potent peptidyl linker for protein cross-linking by microbial transglutaminase from *Streptomyces mobaraensis*. *Bioconjugate Chemistry*, **14**, 351–357.
23. Lin, C.W. and Ting, A.Y. (2006) Transglutaminase-catalyzed site-specific conjugation of small-molecule probes to proteins in vitro and on the surface of living cells. *Journal of the American Chemical Society*, **128**, 4542–4543.
24. Mao, H., Hart, S.A., Schink, A. and Pollok, B.A. (2004) Sortase-mediated protein ligation: a new method for protein

engineering. *Journal of the American Chemical Society*, **126**, 2670–2671.

25 Hodneland, C.D., Lee, Y.S., Min, D.H. and Mrksich, M. (2002) Selective immobilization of proteins to self-assembled monolayers presenting active site-directed capture ligands. *Proceedings of the National Academy of Sciences of the United States of America*, **99**, 5048–5052.

26 Yin, J., Lin, A.J., Golan, D.E. and Walsh, C.T. (2006) Site-specific protein labeling by Sfp phosphopantetheinyl transferase. *Nature Protocols*, **1**, 280–285.

27 Gronemeyer, T., Godin, G. and Johnsson, K. (2005) Adding value to fusion proteins through covalent labeling. *Current Opinion in Biotechnology*, **16**, 453–458.

28 Foley, T.L. and Burkart, M.D. (2007) Site-specific protein modification: advances and applications. *Current Opinion in Chemical Biology*, **11**, 12–19.

29 Chen, I., Choi, Y.A. and Ting, A.Y. (2007) Phage display evolution of a peptide substrate for yeast biotin ligase and application to two-color quantum dot labeling of cell surface proteins. *Journal of the American Chemical Society*, **129**, 6619–6625.

30 Pegg, A.E. (1990) Mammalian O6-alkylguanine-DNA alkyltransferase: regulation and importance in response to alkylating carcinogenic and therapeutic agents. *Cancer Research*, **50**, 6119–6129.

31 Keppler, A., Kindermann, M., Gendreizig, S., Pick, H., Vogel, H. and Johnsson, K. (2004) Labeling of fusion proteins of O6-alkylguanine-DNA alkyltransferase with small molecules *in vivo* and *in vitro*. *Methods*, **32**, 437–444.

32 Keppler, A., Pick, H., Arrivioli, C., Vogel, H. and Johnsson, K. (2004) Labeling of fusion proteins with synthetic fluorophores in live cells. *Proceedings of the National Academy of Sciences of the United States of America*, **101**, 9955–9959.

33 Tanaka, T., Kamiya, N. and Nagamune, T. (2004) Peptidyl linkers for protein heterodimerization catalyzed by microbial transglutaminase. *Bioconjugate Chemistry*, **15**, 491–497.

34 Kamiya, N., Takazawa, T., Tanaka, T., Ueda, H. and Nagamune, T. (2003) Site-specific cross-linking of functional proteins by transglutamination. *Enzyme and Microbial Technology*, **33**, 492–496.

35 Greenberg, C.S., Birckbichler, P.J. and Rice, R.H. (1991) Transglutaminases: multifunctional cross-linking enzymes that stabilize tissues. *FASEB Journal*, **5**, 3071–3077.

36 Lambalot, R.H., Gehring, A.M., Flugel, R.S., Zuber, P., LaCelle, M., Marahiel, M.A., Reid, R., Khosla, C. and Walsh, C.T. (1996) A new enzyme superfamily – the phosphopantetheinyl transferases. *Chemistry & Biology*, **3**, 923–936.

37 Flugel, R.S., Hwangbo, Y., Lambalot, R.H., Cronan, J.E., Jr. and Walsh, C.T. (2000) Holo-(acyl carrier protein) synthase and phosphopantetheinyl transfer in *Escherichia coli*. *Journal of Biological Chemistry*, **275**, 959–968.

38 Gehring, A.M., Lambalot, R.H., Vogel, K.W., Drueckhammer, D.G. and Walsh, C.T. (1997) Ability of Streptomyces spp. acyl carrier proteins and coenzyme A analogs to serve as substrates in vitro for *E. coli* holo-ACP synthase. *Chemistry & Biology*, **4**, 17–24.

39 Yin, J., Straight, P.D., McLoughlin, S.M., Zhou, Z., Lin, A.J., Golan, D.E., Kelleher, N.L., Kolter, R. and Walsh, C.T. (2005) Genetically encoded short peptide tag for versatile protein labeling by Sfp phosphopantetheinyl transferase. *Proceedings of the National Academy of Sciences of the United States of America*, **102**, 15815–15820.

40 Zhou, Z., Cironi, P., Lin, A.J., Xu, Y., Hrvatin, S., Golan, D.E., Silver, P.A., Walsh, C.T. and Yin, J. (2007) Genetically encoded short peptide tags for orthogonal protein labeling by Sfp and AcpS phosphopantetheinyl transferases. *ACS Chemical Biology*, **2**, 337–346.

41 Walsh, C.T., Gehring, A.M., Weinreb, P.H., Quadri, L.E. and Flugel, R.S. (1997)

Post-translational modification of polyketide and nonribosomal peptide synthases. *Current Opinion in Chemical Biology*, **1**, 309–315.

42 Kohsla, C., Gokhale, R.S., Jacobsen, J.R. and Cane, D.E. (1999) Tolerance and Specificity of Polyketide Synthases. *Annual Review of Biochemistry*, **68**, 219–253.

43 Weber, T. and Marahiel, M.A. (2001) Exploring the domain structure of modular nonribosomal peptide synthetases. *Structure*, **9**, R3–R9.

44 Reuter, K., Mofid, M.R., Marahiel, M.A. and Ficner, R. (1999) Crystal structure of the surfactin synthetase-activating enzyme sfp: a prototype of the 4′-phosphopantetheinyl transferase superfamily. *EMBO Journal*, **18**, 6823–6831.

45 Belshaw, P.J., Walsh, C.T. and Stachelhaus, T. (1999) Aminoacyl-CoAs as probes of condensation domain selectivity in nonribosomal peptide synthesis. *Science*, **284**, 486–489.

46 La Clair, J.J., Foley, T.L., Schegg, T.R., Regan, C.M. and Burkart, M.D. (2004) Manipulation of carrier proteins in antibiotic biosynthesis. *Chemistry & Biology*, **11**, 195–201.

47 Yin, J., Liu, F., Li, X. and Walsh, C.T. (2004) Labeling proteins with small molecules by site-specific posttranslational modification. *Journal of the American Chemical Society*, **126**, 7754–7755.

48 Yin, J., Liu, F., Schinke, M., Daly, C. and Walsh, C.T. (2004) Phagemid encoded small molecules for high throughput screening of chemical libraries. *Journal of the American Chemical Society*, **126**, 13570–13571.

49 Chirgadze, N.Y., Briggs, S.L., McAllister, K.A., Fischl, A.S. and Zhao, G. (2000) Crystal structure of Streptococcus pneumoniae acyl carrier protein synthase: an essential enzyme in bacterial fatty acid biosynthesis. *EMBO Journal*, **19**, 5281–5287.

50 Parris, K.D., Lin, L., Tam, A., Mathew, R., Hixon, J., Stahl, M., Fritz, C.C., Seehra, J. and Somers, W.S. (2000) Crystal structures of substrate binding to Bacillus subtilis holo-(acyl carrier protein) synthase reveal a novel trimeric arrangement of molecules resulting in three active sites. *Structure*, **8**, 883–895.

51 George, N., Pick, H., Vogel, H., Johnsson, N. and Johnsson, K. (2004) Specific labeling of cell surface proteins with chemically diverse compounds. *Journal of the American Chemical Society*, **126**, 8896–8897.

52 Vivero-Pol, L., George, N., Krumm, H., Johnsson, K. and Johnsson, N. (2005) Multicolor imaging of cell surface proteins. *Journal of the American Chemical Society*, **127**, 12770–12771.

53 Yin, J., Lin, A.J., Buckett, P.D., Wessling-Resnick, M., Golan, D.E. and Walsh, C.T. (2005) Single-cell FRET imaging of transferrin receptor trafficking dynamics by Sfp-catalyzed, site-specific protein labelling. *Chemistry & Biology*, **12**, 999–1006.

54 Mofid, M.R., Finking, R. and Marahiel, M.A. (2002) Recognition of hybrid peptidyl carrier proteins/acyl carrier proteins in nonribosomal peptide synthetase modules by the 4′-phosphopantetheinyl transferases AcpS and Sfp. *Journal of Biological Chemistry*, **277**, 17023–17031.

55 Quadri, L.E., Weinreb, P.H., Lei, M., Nakano, M.M., Zuber, P. and Walsh, C.T. (1998) Characterization of Sfp, a Bacillus subtilis phosphopantetheinyl transferase for peptidyl carrier protein domains in peptide synthetases. *Biochemistry*, **37**, 1585–1595.

56 Weber, T., Baumgartner, R., Renner, C., Marahiel, M.A. and Holak, T.A. (2000) Solution structure of PCP, a prototype for the peptidyl carrier domains of modular peptide synthetases. *Structure*, **8**, 407–418.

57 Holak, T.A., Kearsley, S.K., Kim, Y. and Prestegard, J.H. (1988) Three-dimensional structure of acyl carrier protein determined by NMR pseudoenergy and distance geometry calculations. *Biochemistry*, **27**, 6135–6142.

58 Koglin, A., Mofid, M.R., Lohr, F., Schafer, B., Rogov, V.V., Blum, M.M., Mittag, T., Marahiel, M.A., Bernhard, F. and Dotsch, V. (2006) Conformational switches modulate protein interactions in peptide antibiotic synthetases. *Science*, **312**, 273–276.

59 Li, Q., Khosla, C., Puglisi, J.D. and Liu, C.W. (2003) Solution structure and backbone dynamics of the holo form of the frenolicin acyl carrier protein. *Biochemistry*, **42**, 4648–4657.

60 Meyer, B.H., Segura, J.M., Martinez, K.L., Hovius, R., George, N., Johnsson, K. and Vogel, H. (2006) FRET imaging reveals that functional neurokinin-1 receptors are monomeric and reside in membrane microdomains of live cells. *Proceedings of the National Academy of Sciences of the United States of America*, **103**, 2138–2143.

61 Marahiel, M.A., Stachelhaus, T. and Mootz, H.D. (1997) Modular peptide synthetases involved in nonribosomal peptide synthesis. *Chemical Reviews*, **97**, 2651–2674.

62 Barbas, C.F., 3rd, Kang, A.S., Lerner, R.A. and Benkovic, S.J. (1991) Assembly of combinatorial antibody libraries on phage surfaces: the gene III site. *Proceedings of the National Academy of Sciences of the United States of America*, **88**, 7978–7982.

63 Vitali, F., Zerbe, K. and Robinson, J.A. (2003) Production of vancomycin aglycone conjugated to a peptide carrier domain derived from a biosynthetic non-ribosomal peptide synthetase. *ChemComm*, 2718–2719.

64 Clarke, K.M., Mercer, A.C., La Clair, J.J. and Burkart, M.D. (2005) In vivo reporter labelling of proteins via metabolic delivery of coenzyme A analogues. *Journal of the American Chemical Society*, **127**, 11234–11235.

65 Schultz, C. (2003) Prodrugs of biologically active phosphate esters. *Bioorganic and Medicinal Chemistry*, **11**, 885–898.

8
Bioorthogonal Chemical Transformations in Proteins by an Expanded Genetic Code

Birgit Wiltschi and Nediljko Budisa

8.1
Introduction

Living cells produce hundreds of thousands of proteins carrying out stupendously manifold biological functions. Nevertheless, the inventory of side chains of amino acids incorporated into proteins during translation at the ribosome is quite limited. All living beings use the same 20 canonical amino acids as building blocks for protein biosynthesis (Figure 8.1), and only recently has this repertoire been expanded to 22 members including selenocysteine [1] and pyrrolysine [2]. This raises the question how nature accomplishes biological processes requiring functionalities that are not encoded by the genetic code: The diversity of the cellular protein inventory, that is, the proteome, can be enormously expanded by covalent modifications that are introduced into the proteins at one or more sites after translation [3, 4]. This post-translational processing comprises a battery of specific enzyme-catalyzed modifications on the amino acid side chains or backbones.

Although naturally occurring proteome diversity is vast, it does not (or only unsatisfactorily) offer biotechnologically interesting options, such as the introduction of atomic markers (isotopes or heavy atoms), redox-sensitive groups, fluorescent moieties, glycosylated amino acids, reporter groups or therapeutic compounds at well defined positions. The regioselective decoration of native proteins with non-natural structural elements, for example, for the covalent attachment of other biomolecules to form protein conjugates, for site-selective immobilization of proteins to solid supports, or the addition of hydrophobic residues necessary for membrane anchoring, presents an eminently attractive goal. Hitherto unsuspected possibilities for the tailored modification of protein function would arise which cannot be implemented by current biological or artificial methods.

The chemistry required for manipulation of native proteins is available in quite a variety [6–11]. However, the severest shortcoming of these labeling reactions is their non-specificity: Either they can occur simultaneously on different amino acid residues, resulting in heterogeneous labelling, or the simultaneous labeling of those

Probes and Tags to Study Biomolecular Function. Lawrence W. Miller (Ed.)
Copyright © 2008 WILEY-VCH Verlag GmbH & Co. KGaA, Weinheim
ISBN: 978-3-527-31566-6

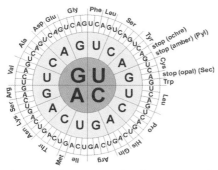

Figure 8.1 The genetic code. The 20 canonical amino acids are shown in three-letter code. Selenocysteine (Sec) and pyrrolysine (Pyl) have been added to the genetic code recently. In contrast to the other canonical amino acids, the translation of Sec and Pyl is context dependent (reviewed in [5]).

few functional groups that can be modified specifically, such as sulfhydryl groups [12], cannot be avoided if a protein has many such groups [13].

Briefly, a single site-specific labeling method is the primary prerequisite to regioselectively address target epitopes in folded proteins. Further, if the desired chemical transformation includes the reaction between two chemical moieties they should react exclusively with each other. Other components in the reaction mix should not interfere, that is, the reaction should be bioorthogonal. Keto-, aldehyde-, alkyne- or azido-groups offer the bioorthogonality necessary for chemical transformations of native proteins. Although the genetic code does not include these functionalities, several techniques have been developed to add them to it. This chapter reviews different approaches to exploit genetic code engineering with bioorthogonal groups for chemical transformations of native proteins (Figure 8.2).

8.2
Chemical Transformations at the Protein N-terminus Classical Approaches

8.2.1
Biomimetic Transamination

The free α-amino group of the N-terminal amino acid of a protein is an ideal site for specific ligation with a marker moiety, because this group is usually not essential for protein function [14].

Transamination of the free α-amino group at the N-terminal amino acid represents a classical site-specific chemical transformation of a protein. Thereby, the amino group reacts with glyoxylic acid or pyridoxal-5′-phosphate (PLP), resulting in the transamination of the N-terminal residue to form a 2-oxoacyl group [15]

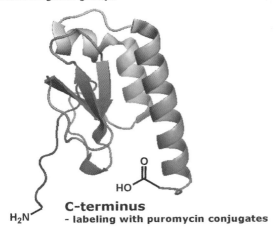

Figure 8.2 Overview of the chemical transformations in proteins that will be discussed in this chapter.

(Figure 8.3A). Although both compounds yield the same product, transamination with PLP is preferable with native proteins as it proceeds in aqueous solution under milder conditions [16] than the transamination with glyoxylic acid (Figure 8.3A). During the transamination reaction an intermediate is formed that involves the participation of the adjacent peptide bond. This ensures that only the N-terminal amino group is converted, while the internal amino groups on lysine residues are not modified [17].

Only recently, Gilmore and coworkers performed a detailed study of the reaction conditions for N-terminal transamination of model peptides and proteins. They found that the reaction readily proceeded with PLP within 2 h at 37 °C in standard aqueous buffers. It did not require the presence of divalent cations or denaturing organic co-solvents. Transamination efficiencies depended on the N-terminal residue and ranged between 30 and 80%. Proteins with N-terminal methionine residues are well compatible with this technique. However, proteins possessing N-terminal serine, threonine, cysteine, and tryptophan residues are most probably incompatible because of known side reactions with aldehydes. Furthermore, N-terminal proline residues will be unreactive (see [16] and references therein).

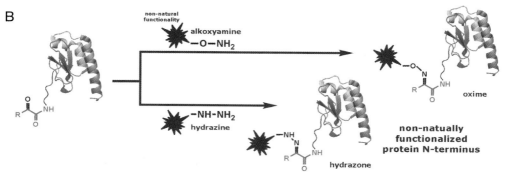

Figure 8.3 Transamination of N-terminal α-amino groups and subsequent coupling reactions (A) Transamination reaction with glyoxylic acid (upper part) and pyridoxal-5′-phosphate (PLP) (lower part). Note the relatively harsh reaction conditions with glyoxylic acid in comparison to PLP. (B) Coupling reaction with hydrazines or alkoxyamines to form hydrazones or oximes, respectively.

N-terminal transamination introduces a uniquely reactive ketone or aldehyde group in a single location. The carbonyl group reacts with compounds containing a hydrazine or alkoxyamine group to form a hydrazone or oxime, respectively, with no cross-reactivity (Figure 8.3B). Aside from the limitations imposed by particular N-terminal residues, biomimetic transamination is a simple strategy that has the potential to introduce virtually any functional group on a wide range of protein substrates with a free N-terminal α-amino group. Site-directed mutagenesis to direct the modification to the desired position is not necessary.

Biomimetic transamination together with hydrazine- or alkoxyamine coupling was successfully used for the N-terminal labeling of enzymes and antibodies with fluorophores or polyethylene glycol ([16, 18]). Dong *et al.* designed a polarity-sensitive fluorescent probe with a terminal hydrazine group and covalently coupled it to the transaminated N-terminus of α-lactalbumin. With the conjugated probe they were able to detect the local polarity changes of the N-terminal domain in both native and heat-denatured α-lactalbumin [13].

8.2.2
Enzymatic Modification of the N-terminus

The modification of the protein N-terminus using enzymes benefits from the high specificity and efficiency of enzymatic reactions under physiological conditions. The technique should be generally applicable to a wide variety of proteins, provided that they have flexible or exposed N-terminal regions. A major drawback, however, resides in the inevitable recombinant engineering of the target proteins for them to serve as substrates for the enzyme.

Lewinska et al. reported a method to regioselectively elongate the N-terminus of proteins with non-natural moieties by kinetically controlled reverse proteolysis (Figure 8.4) [19]. To achieve this, the authors used IgA-protease which specifically recognizes a stretch of seven amino acids [20] and contains the scissile peptide bond (↓) in the middle of the recognition sequence: (Ser, Pro)-(Arg, Thr)-Pro-Pro ↓ (Ala, Gly, Ser)-Pro-(Ser, Trp, Tyr) (see [19] and references therein). This specificity determinant rarely occurs in globular proteins, thus minimizing the risk of unspecific hydrolysis of protein substrates. Since all enzymes catalyze reversible chemical reactions, IgA-protease will not only cleave the determinant but, under tightly controlled conditions, also covalently conjugate two peptidic fragments carrying the matching halves of the recognition sequence. The H_2N-(Ala, Gly, Ser)-Pro-(Ser, Trp, Tyr) sequence of the determinant, which acts as the amino nucleophile, is engineered onto the N-terminus of the target protein (Figure 8.4). IgA protease can conjugate this fragment with a tetrapeptidic acyl donor of the sequence X-(Ser, Pro)-(Arg, Thr)-Pro-Pro-OMe, where X signifies a desired (non-natural) residue and OMe is a carboxy-terminal ester group (Figure 8.4). Typically, the acyl donor is synthetically accessible, and, within generous limits, the chemical nature of the non-natural group X is uncritical to the success of the reaction [19]. The kinetic control refers to the exact tuning of the quench process that will stop the enzyme catalysis when the yield of the coupling product reaches a maximum.

In an alternative enzymatic approach, the groups of Hasegawa and Sisido used L/F-tRNA-protein transferase for the modification of the protein N-terminus. L/F-tRNA-protein transferase catalyzes the transfer of some hydrophobic amino acids (such as phenylalanine, leucine or methionine) from aminoacylated tRNA to a protein with N-terminal lysine or arginine residues [21]. Usually protein biosynthesis starts with methionine, and N-terminal methionine excision is responsible for the diversity of N-terminal amino acids in proteins [22]. However, in order to expose the N-terminal lysine or arginine residues necessary for proper action of L/F-tRNA-protein transferase, again, the target protein has to be genetically engineered. Kuno and coworkers introduced the pelB signal peptide succeeded by lysine into their model protein (Figure 8.5). Expression of this fusion protein in E. coli and concomitant intracellular cleavage of the pelB signal peptide resulted in mature protein with an N-terminal lysine. For the label, transfer RNA for phenylalanine (tRNA^{Phe}) was aminoacylated with

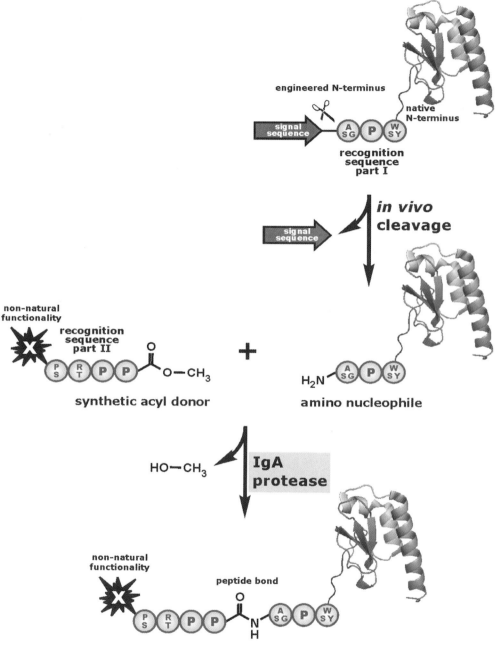

radioactive phenylalanine (F*) by phenylalanyl-tRNA synthetase (PheRS). Subsequently, F* from F*-tRNA^{Phe} was transferred to the N-terminal lysine of the target protein by purified recombinant L/F-tRNA-protein transferase (Figure 8.5A) [23].

Taki *et al.* developed the method further and used L/F-tRNA-protein transferase to regioselectively attach non-canonical amino acids with bioorthogonal reactive groups (Figure 8.5b) to N-terminal lysine residues [24, 25]. Considering the heterogeneous chemical structures of these non-canonical amino acids (Figure 8.5B), the substrate flexibility of the L/F-tRNA-protein transferase is strikingly broad, yet finite since 9-anthrylalanine was unreactive [25]. In order to avoid inadequate recognition of the non-canonical amino acids by the PheRS, Taki and coworkers charged them onto tRNA^{Phe} by chemical aminoacylation according to Hecht [26]. Aminoacyl transfer by L/F-tRNA-protein transferase is highly selective and quite robust since the non-canonical amino acids were transferred successfully to the N-termini of target peptides as well as proteins in the presence of other peptides and a crude protein mixture, respectively [24].

The newly introduced reactive bioorthogonal groups at the protein N-terminus facilitate derivatization with non-natural functionalities, for example, fluorophores or biotin (Figure 8.5C). Carbonyl groups, for example, of *p*-acetyl-L-phenylalanine, can be conjugated with hydrazines or alkoxyamines to synthesize labeled protein via a hydrazone or oxime linkage, respectively. Azido groups as in *p*-azido-L-phenylalanine undergo Huisgen [3 + 2] cycloaddition exclusively with alkyne moieties to form a stable triazole conjugate [24]. The advantage of using bioorthogonal functional groups is that even such bulky tags can be conjugated to the protein N-terminus that would be unamenable to enzymatic transfer by L/F-tRNA-protein transferase.

8.3
Chemical Transformations Using an Expanded Genetic Code

8.3.1
The Genetic Code and Its Expansion

The labeling methods described above regioselectively modify the N-terminus of a mature protein in its native folded state. Alternatively, one can incorporate the uniquely addressable moiety at the stage of protein synthesis. Since bioorthogonal groups are usually not part of the canonical amino acid repertoire (Figure 8.1), their introduction into polypeptides inevitably requires the expansion of the genetic code.

Figure 8.4 Chemo-enzymatic elongation of the protein N-terminus. Typically, the N-terminus of the target protein is genetically engineered by introduction of the recognition sequence (Ala, Gly, Ser)-Pro-(Ser, Trp, Tyr) at the junction of a signal sequence and the native N-terminus of the protein. Upon enzymatic cleavage (✂) of the signal sequence, the recognition sequence is exposed (right) and can be conjugated to the synthetic acyl donor (left) by IgA-protease. The final protein conjugate carries the non-natural functionality introduced by X.

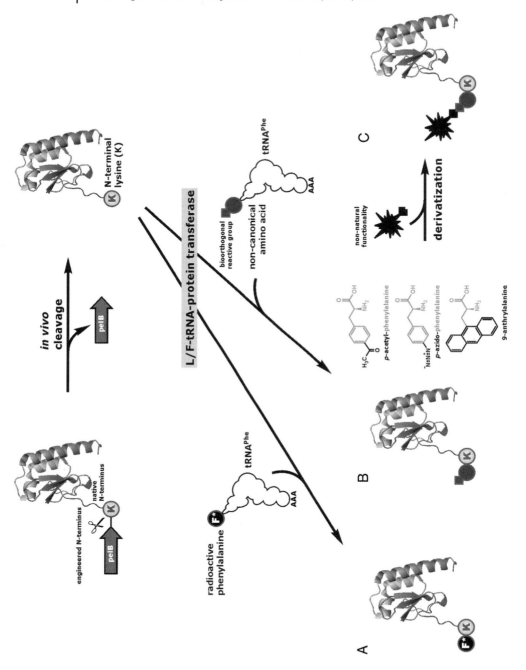

All living organisms store genetic information as a sequence of four nucleotides in their DNA molecules. In order to execute the genetic information, the DNA is transcribed to messenger RNA (mRNA) and the information is translated into a polypeptide sequence at the ribosome. The precision of the information transfer from nucleic acids to proteins is governed by a set of rules, that is, the genetic code (Figure 8.1), that strictly links a linear nucleotide sequence to a linear sequence of amino acids. The genetic code operates with tri-nucleotide units (codons) each of which codes for a single amino acid or translation termination signal (Figure 8.1). There are 64 possible three-base combinations of the four nucleotides, enough to encode 64 amino acids. However, as already outlined in the introductory section, all life forms on Earth use only a standard set of 20 canonical amino acids to biosynthesize their proteins.

Amino acids are provided for ribosomal protein synthesis in the form of aminoacyl-tRNAs. The transfer RNA (tRNA) molecule acts as the "adaptor" that transfers the genetic information from the nucleic acid (mRNA) to the polypeptide [27]. The protein translation process encompasses two distinct molecular recognition events: the interaction of the aminoacyl-tRNA anticodon with the correct codon in the mRNA on the ribosome and the specific pairing of the tRNA anticodon with its pertinent amino acid by action of the aminoacyl-tRNA synthetase (AARS). Thus, the specific covalent attachment of an amino acid to its cognate tRNA, which interprets the genetic code, is controlled by the substrate specificity and selectivity of the AARS. However, the AARSs demonstrate catalytic promiscuity [28, 29] and the ribosome is surprisingly tolerant with respect to the chemical nature of amino acid side chains [30]. Consequently, the genetic code is malleable and quite a number of non-canonical amino acids can be smuggled into the canonical repertoire.

8.3.2
Cell-free N-terminal Labeling with Modified Initiator tRNA

Cell-free protein labeling techniques exploit the fact that as soon as an aminoacylated tRNA is accepted by the ribosome, the amino acid is incorporated into the nascent polypeptide chain without further proofreading. Ribosomal protein synthesis starts with methionine charged onto a tRNA that specifically recognizes AUG start codons, the so-called initiator tRNA, that is, $tRNA_i$ in eukaryotic cells and $tRNA^{fMet}$ in prokaryotes. Thus, for N-terminal labeling, the initiator tRNA must be mischarged with an amino acid carrying the desired non-natural functionality.

The Rothschild group successfully incorporated non-natural functionalities at the N-terminus of nascent proteins translated in an *E. coli*-based *in vitro* coupled

Figure 8.5 Enzymatic modification of the protein N-terminus with L/F-tRNA-protein transferase. (A) Labeling with radioactive phenylalanine (F*). (B) Conjugation of non-canonical amino acids with bioorthogonal reactive groups to the N-terminus of the target protein. (C) Subsequent derivatization of the bioorthogonal groups with non-natural functionalities.

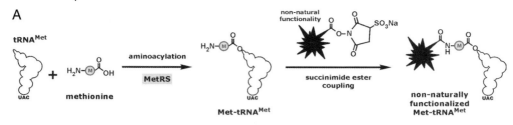

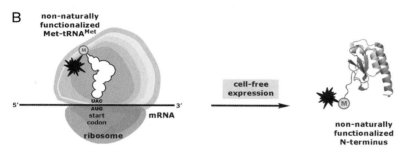

Figure 8.6 N-terminal labeling using modified initiator tRNA^{fMet} in an *E. coli*-based *in vitro* protein expression system. (A) Chemical modification of methionine directly on the tRNA^{fMet}. First, methionine is enzymatically charged onto tRNA^{fMet} by methionyl-tRNA synthetase (MetRS). The α-amino group of methionine is subsequently chemically conjugated to the desired non-natural functionality, for example, a fluorophore or biotin. (B) Cell-free incorporation of the methionine conjugate into a nascent polypeptide in response to an AUG start codon. The N-terminus of the resulting protein carries the desired non-natural functionality.

transcription/translation system (Figure 8.6) [31, 32]. To achieve this, *E. coli* wild-type tRNA^{fMet} was enzymatically charged with methionine to yield methionyl-tRNA^{fMet} (Met-tRNA^{fMet}). Subsequently, the methionine α-amino group was chemically conjugated with the fluorescent dye BODIPY or biotin directly on the tRNA^{fMet} using succinimide ester coupling chemistry [31] (Figure 8.6A). The misaminoacylated BODIPY- and biotin-Met-tRNA^{fMet} were then introduced into cell-free protein synthesis (Figure 8.6B) and the resulting proteins analyzed by detection of fluorescent product bands on a protein gel or after immunoblotting with streptavidin–alkaline phosphatase, respectively [31, 32].

Using the same method, Taki *et al.* were able to biotinylate the N-terminus of green fluorescent protein without affecting the native function of the protein [33]. Tsalkova *et al.* used coumarin-Met-tRNA^{fMet} to study the influence of a hydrophobic probe at the N-termini of bacterial chloramphenicol acetyltransferase and bovine mitochondrial rhodanese on translational pausing in an *E. coli in vitro* coupled transcription/translation system [34].

Although N-terminal labeling with misaminoacylated initiator tRNA is feasible, the labeling efficiency is rather low due to competition between the exogenously added misaminoacylated initiator tRNA^{fMet} and endogenous formyl-Met-tRNA^{fMet} in the *E. coli* cell-free translation system. Olejnik *et al.* achieved only 1–2% labeling efficiency, while Taki and coworkers found 10–20%, yet were unable to increase the efficiency further after removal of non-labeled Met-tRNA^{fMet} and free tRNA^{fMet} from the modification reaction mixture [35].

An explanation for the competition reaction might arise from the translation initiation process itself. During initiation of bacterial protein synthesis, initiation factor 2 (IF-2) binds formyl-Met-tRNA^{fMet} and positions it over the start codon of the mRNA in the ribosomal P site (reviewed in [36, 37]). McIntosh and coworkers studied in detail the ability of various fluorescent α-amino group derivatives of Met-tRNA^{fMet} to function in polypeptide synthesis in the *E. coli* cell-free expression system. Their results indicate that fluorophore-Met-tRNA^{fMet} species can, and obviously do, participate in IF-2-mediated initiation reactions. Reduced interaction with IF-2, however, and inefficient binding to ribosomes, but not misalignment within the P site, most likely contribute to the low incorporation of N-terminal probes into polypeptides [38].

8.3.3
Cell-free N-terminal Labeling with Suppressor Initiator tRNA

In order to overcome the competition problem, a simplified, reconstituted, AARS-free, pure translation system [39] containing exclusively misacylated tRNA^{fMet} could be used.

The Rothschild group reported an alternative system utilizing a mutant mRNA template with a UAG amber codon in place of a normal AUG start codon and a complementary misaminoacylated initiator suppressor tRNA^{fMet}(CUA) capable of initiating protein synthesis from the UAG codon [14, 31]. The studies of RajBhandary and coworkers demonstrated that initiator tRNAs with altered anticodons are capable of initiating protein synthesis with amino acids other than methionine [40]. In contrast to suppression of termination codons, initiation suppression has the advantage of no competition with release factors and the initiator suppressor does not suppress termination codons. Depending on the protein, N-terminal labeling efficiencies between 27 and 67% were observed with different fluorophore-amino acid conjugates [14].

Taken together, cell-free N-terminal labeling with misaminoacylated initiator tRNAs is versatile and specific. However, the product yields are rather low. So far no *in vivo* methodologies for specific labeling of the protein N-terminus have been developed.

8.4
Modifications Internal to Proteins

In principle, exchange of any of the 20 canonical amino acids by misaminoacylation of the cognate tRNA with a non-canonical analog is possible in cell-free protein expression systems [41]. In this way, ribosomal translation is reprogrammed for the

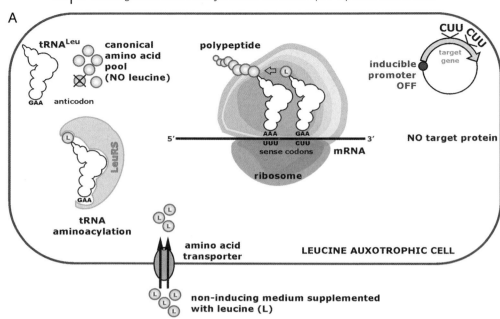
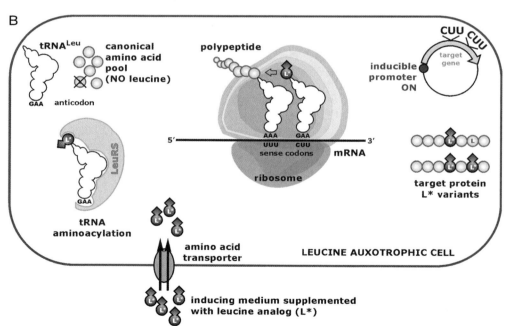

introduction of one or more non-canonical amino acids throughout the polypeptide chain, either residue-specifically, that is, in response to any of the sense codons, or site-specifically, if the specific codon appears at a single site in the target protein or if translation termination codons are read through by misaminoacylated suppressor tRNAs. While the diversity of the amino acid side chains appears to be limited only by the flexibility of the ribosome, a major drawback of *in vitro* protein expression is the very low product yield.

In order to surmount this shortcoming the *in vivo* incorporation of non-canonical amino acids into target proteins has been explored. Two main approaches have been developed for *in vivo* expansion of the genetic code by reprogramming protein translation. The first technique is called selective pressure incorporation (SPI; Figure 8.7) and allows residue-specific incorporation of a non-canonical amino acid into the target protein. It exploits the already mentioned catalytic promiscuity of the AARS. In this way, non-canonical amino acids that are structurally similar to their canonical analogs can be smuggled into ribosomal protein synthesis. However, the canonical amino acid is always the preferred substrate for the AARS. For efficient incorporation of the non-canonical amino acid into the target protein the expression host must be auxotrophic, that is, biosynthesis deficient for the cognate canonical amino acid. Moreover, a strongly inducible expression system for the target protein is necessary. The cells are first cultured under non-inducing growth conditions in the presence of the canonical amino acid (Figure 8.7A). After accumulation of cell mass, the cells are transferred to an inducing medium containing the non-canonical amino acid (Figure 8.7B). Most non-canonical amino acids do not support cell growth, nevertheless, ribosomal protein biosynthesis is still active. Due to the strong induction of target protein expression, *de novo* protein synthesis predominantly produces target protein in which certain canonical amino acid residues are statistically replaced by the non-canonical analog. There is never 100% replacement efficiency, most probably because of the intracellular release of canonical amino acids by bulk protein turn-over.

Alternatively, stop codon suppression (SCS [42]; Figure 8.8) facilitates the site-specific incorporation of non-canonical amino acids into target proteins in response to translation termination signals [43]. In order to achieve this, an orthogonal mutant AARS/suppressor tRNA pair must be evolved which is specific for the non-canonical

---

Figure 8.7 Selective pressure incorporation of non-canonical amino acids. For example, a leucine analog is to be incorporated into a target protein. (A) Leucine-auxotrophic cells are cultivated under non-inducing conditions in the presence of leucine (L). Since the inducible promoter is shut OFF no target protein is produced. All translated cellular proteins contain leucine. (B) After accumulation of cell mass the cells are transferred to an inducing medium containing a non-canonical leucine analog (L*). Usually, the cells stop growing in the presence of non-canonical amino acids but translation is still active. Due to the catalytic promiscuity of the leucyl-tRNA synthetase (LeuRS) the L* is charged onto tRNA^{Leu} and is incorporated into the target protein whose expression is now ON. The strong promoter drives *de novo* protein synthesis predominantly of the target protein. L* is statistically incorporated in response to leucine-encoding CUU codons. The incorporation efficiency depends on the intracellular release of leucine by the turn-over of bulk protein.

amino acid to be incorporated (Figure 8.8A). Here, orthogonality means that the mutant AARS must not charge any of the intracellular tRNAs with the non-canonical amino acid nor may the suppressor tRNA be charged with a canonical amino acid by any of the cellular AARSs. The orthogonal AARS/tRNA pair is constitutively expressed in the expression host, together with the target protein whose coding

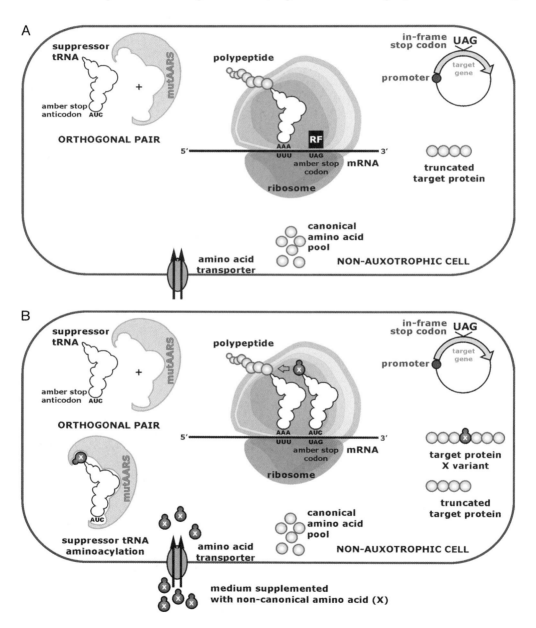

sequence carries a stop codon at the position where the non-canonical amino acid is desired in the translated protein. In this case, the host strain need not be auxotrophic for the specific canonical amino acid since the orthogonal AARS/tRNA pair requires the presence of the non-canonical amino acid for proper function (Figure 8.8B).

A plethora of non-natural functionalities has been incorporated into target proteins using either of the two techniques [43, 44]. Although a comprehensive list would go far beyond the scope of this review the tailoring of protein fluorescence deserves a short remark. Trp is the main source of fluorescence and absorbance in proteins. The spectral properties of non-canonical Trp analogs are even more diverse as they include quenching, spectra shifts and electron-donating or -withdrawing properties. The most outstanding analogs are aza- and amino-tryptophans which exhibit strong pH-dependent fluorescence as well as significant red or blue shifts in fluorescence emission in comparison to Trp. For instance, substitution of Trp with 4-amino-tryptophan in the chromophore of a novel "gold" class of autofluorescent proteins results in a 100 nm Stokes shift of the fluorescence maximum [45].

However, some non-canonical amino acids with interesting side chain functionalities are not translationally active in living cells. Proteins can still be furnished with these functions by the co-translational incorporation of bioorthogonal moieties which allow chemical transformation after protein translation. Non-canonical amino acids with terminal azido- or alkyne side chains (refer to examples in Figure 8.9A) are especially useful since these form covalent triazole products exclusively with terminal alkyne or azido-conjugates, respectively, in a Cu(I)-catalyzed Huisgen [3 + 2] cycloaddition (Sharpless–Meldal "click" chemistry [46–48]; Figure 8.9B). For instance, newly synthesized proteins in living cells were fluorescently labeled by co-translational SPI of homopropargylglycine (Figure 8.9A) followed by post-translational derivatization ("clicking") with 3-azido-7-hydroxycoumarin [49, 50]. Schultz's group incorporated aromatic alkyne and azido amino acids in response to an amber stop codon and successfully derivatized them with azido- and alkyne-fluorophore conjugates, respectively [51]. Alternatively, azido-groups can be covalently linked to phosphine-conjugates by the Staudinger–Bertozzi ligation [52–54] (Figure 8.9C, upper part) or traceless Staudinger ligation [55, 56] (Figure 8.9C, lower part). The

Figure 8.8 Amber stop codon suppression with a non-canonical amino acid. (A) Non-auxotrophic cells are cultivated in the absence of the non-canonical amino acid (X). An orthogonal suppressor tRNA/mutated AARS (mutAARS) pair is constitutively expressed. As long as X is absent from the medium, the orthogonal pair remains inactive. Thus, the in-frame stop codon in the target protein is recognized by a release factor (RF) which terminates translation and only truncated target protein is expressed. (B) After accumulation of cell mass the cells are transferred to a medium containing X and now mutAARS exclusively charges the suppressor tRNA with X. Aminoacylated suppressor tRNA competes with RF for the stop codon and X is incorporated into the target protein in response to the amber stop codon. Depending on the competition efficiency, truncated protein can be produced as well. Unspecific stop codon suppression in cellular proteins is negligible since the target protein is the main product of *de novo* protein synthesis.

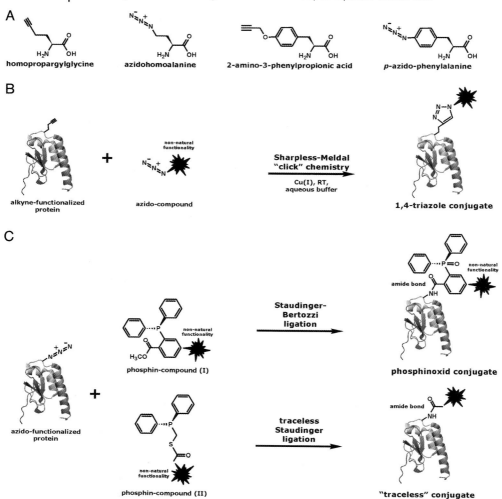

Figure 8.9 Post-translational modifications with bioorthogonal azido or alkyne groups. (A) Non-canonical amino acids with alkyne (homopropargylglycine; 2-amino-3-phenylpropionic acid) and azido (azidohomoalanine; p-azido-phenylalanine) bioorthogonal groups. (B) Cu(I)-catalyzed Huisgen [3 + 2] cycloaddition (Sharpless-Meldal "click" chemistry). (C) Staudinger–Bertozzi (upper) and traceless Staudinger (lower) ligations.

group of Bertozzi demonstrated that cell surfaces in living animals can be chemically remodeled with a specific antigen by the aid of the Staudinger ligation [57].

The diversity of chemical transformations by an expanded genetic code appears infinite. It has to be kept in mind, though, that labeling of internal amino acid residues can interfere with protein function.

8.5
Chemical Transformations at the Protein C-terminus

Protein N- and C-terminal regions usually have large flexibility and tend to be located far away from the active site. Similar to modifications at the protein N-terminus, carboxy-terminal labeling should be less perturbing for protein function than chemical modification of internal residues [58].

The antibiotic puromycin (Figure 8.10A) acts as an inhibitor of protein biosynthesis [59–62] in prokaryotes and eukaryotes [63]. Since the structure of puromycin mimics the 3′-end of tyrosyl-tRNA^{Tyr} [63, 64], it competes with aminoacylated tRNA for entry at the acceptor site (A site) of the ribosome. The growing polypeptide chain is transferred to the α-amino group of the pseudo-amino acid of puromycin (Figure 8.10A). This interrupts the normal reaction of peptide bond formation and results in non-specific linkage of puromycin to the polypeptide chain and subsequent premature elongation termination (Figure 8.10B).

Only recently Miyamoto-Sato et al. [65] observed that in in vitro translation, at very low concentrations (e.g. 0.04 µM) puromycin does not compete effectively with aminoacyl-tRNA and non-specific translation inhibition does not occur (Figure 8.10C). Apparently, puromycin at sufficiently low concentrations gets a chance to be bonded to proteins only at a stop codon, where it does not need to compete with aminoacyl-tRNA but with far less abundant release factors. Consequently, the antibiotic is specifically linked to the C-terminus of the polypeptide chain at the stop codon in the process of normal termination of protein synthesis [65]. This represents a special expansion of the genetic code since the meaning of a stop codon is changed from "terminate translation" to "link puromycin."

Based on the above described finding, the same group developed a new C-terminus-specific protein labeling method with puromycin in an in vitro translation system (Figure 8.11A). In order to improve the incorporation at the protein C-terminus, an mRNA without a stop codon was used. The ribosome stalls at the 3′-end of the truncated mRNA without a stop codon because aminoacyl-tRNAs or release factors cannot associate with the A site of the ribosome and the polypeptide chain cannot be elongated nor released. Thus, only puromycin enters the A site of the ribosome, and bonds to the C-terminus of the polypeptide chain.

Using this approach, Nemoto et al. successfully incorporated a fluorescent puromycin analog (Figure 8.11B) at the protein C-termini of different model proteins [66]. The labeling efficiency with fluorescein-puromycin in the µM range varied between 50 and 95% and depended on the model protein and the cell-free protein synthesis system used [66]. Doi et al. evaluated the general utility of this labeling method by employing puromycin analogs linked to various fluorophores through a deoxycytidylic acid linker (Figure 8.11C) as C-terminally specific fluorescent reagents [67]. Kobayashi and coworkers reported that the low yields of labeled proteins were mainly dependent on the C-terminal amino acid sequence. They showed that the short peptide tag sequence, RGAA, at the C-terminus produced a high yield of labeled protein and high labeling efficiency with all of the cell-free translation systems tested [68].

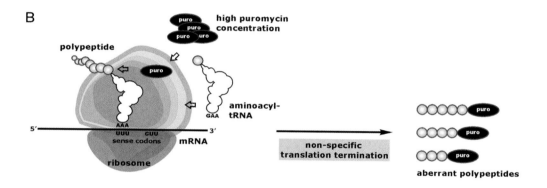

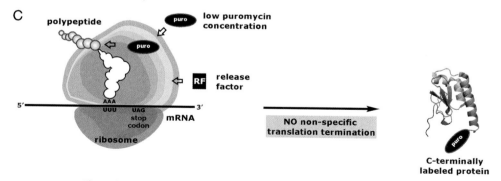

Figure 8.10 Puromycin and its mode of action. (A) Chemical structure of puromycin in comparison to aminoacylated tRNA^Tyr. (B) Inhibition of translation by high puromycin concentrations. The antibiotic binds non-specifically to the nascent polypeptide in competition with aminoacyl-tRNA. (C) At low concentrations, puromycin cannot compete with aminoacyl-tRNAs to inhibit translation but competes with far less abundant release factors for stop codons. Specific bonding of the antibiotic to the C-terminus of full-length protein occurs.

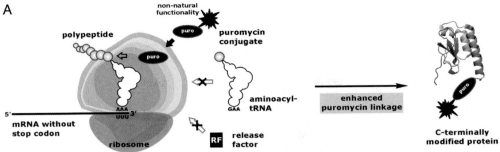

Figure 8.11 C-terminal protein transformations using puromycin conjugates. (A) Incorporation of a puromycin conjugate at the C-terminus of the target protein is enhanced if an mRNA without a stop codon is used. (B) Fluorescein conjugate of puromycin. (C) Puromycin analog linked to fluorescein via a deoxycytidylic acid linker. (D) Structure of 2′-deoxycytidylyl-(3′ → 5′)-puromycin modified by a linker carrying a bioorthogonal azide group [72].

Rapid and reliable fluorescence labeling of (complex) protein samples without loss of function is crucial for fluorescence-based high throughput analysis of protein interactions, such as microarray technologies. The C-terminal incorporation of fluorescent puromycin analogs appears especially attractive for the large-scale labeling of proteins as protein expression and labeling steps are synchronized and a purification step before labeling is unnecessary. Recently, protein–protein and protein–DNA interactions have been analyzed successfully with carboxy-terminally fluorescence labeled proteins [67, 69–71]. In these experiments, translational C-terminal labeling proved to be superior over conventional chemical modification of, for example, internal lysines, with respect to the preservation of protein function [67].

The latest progression of the puromycin method was introduced by Humenik and coworkers. They incorporated bioorthogonal functional groups into the C-termini of proteins by using azide-functionalized puromycin analogs (Figure 8.11D) in a cell-free protein expression system. In combination with "click" chemistry, efficient site-specific conjugation with 5′-alkyne oligonucleotides was achieved and an extension of this conjugation method to fluorescent labels, saccharides, or peptides is likely to be possible [72].

In principle, a fluorescent or biotinylated variant of puromycin should be functional in protein synthesis *in vivo* if it is able to enter the cells in a non-destructive fashion. Starck *et al.* demonstrated that a variety of puromycin conjugates easily enter cells and covalently label newly synthesized proteins at their C-termini. In this way, selective and site-specific *in vivo* labeling of newly synthesized proteins facilitates direct monitoring of protein expression and provides the potential for both spatial and temporal resolution [73].

8.6
Summary and Outlook

Classical transformation chemistries for protein modification suffer from their nonspecificity. They lead to random chemical modification and functional heterogeneity of the products. Novel approaches combine site-specific co-translational incorporation of bioorthogonal groups into proteins with their chemoselective post-translational derivatization. Protein translation reprogrammed with non-canonical amino acid analogs carrying bioorthogonal azide or alkyne moieties facilitates their site-specific incorporation at the protein N- or C-termini or internal to the polypeptide chain. Subsequent bioorthogonal chemical reactions, such as ketone/aldehyde-hydrazine reactions, the Staudinger ligation and Cu(I)-catalyzed Huisgen [3 + 2] cycloaddition, allow regio- and chemoselective modification of the labeled proteins without cross reactivity.

Although we have only just begun to explore the potential of genetic code engineering it is already evident that this technique starts an entirely new era in protein modification that offers virtually limitless (biotechnological) options.

References

1. Böck, A., Forchhammer, K., Heider, J. and Baron, C. (1991) Selenoprotein synthesis: an expansion of the genetic code. *Trends in Biochemical Sciences*, **16**, 463–467.
2. Hao, B., Gong, W., Ferguson, T.K., James, C.M., Krzycki, J.A. and Chan, M.K. (2002) A New UAG-Encoded Residue in the Structure of a Methanogen Methyltransferase. *Science*, **296**, 1462–1466.
3. Walsh, C.T. (2006) *Posttranslational Modifications of Proteins: Expanding Nature's Inventory*, Roberts & Co., Englewood, Colorado.
4. Walsh, C.T., Garneau-Tsodikova, S. and Gatto, G.J.J. (2005) Protein Posttranslational Modifications: The Chemistry of Proteome Diversifications. *Angewandte Chemie-International Edition in English*, **44**, 7342–7372.
5. Wiltschi, B. and Budisa, N. (2007) Natural history and experimental evolution of the genetic code. *Applied Microbiology and Biotechnology*, **74**, 739–753.
6. Prescher, J.A. and Bertozzi, C.R. (2005) Chemistry in living systems. *Nature Chemical Biology*, **1**, 13–21.
7. Offord, R.E. and Gaertner, H.F. (2003) Semisynthesis, in *Hobuben-Weyl, Methods of Organic Chemistry: Synthesis of Peptides and Peptidomimetics* (eds M. Goodman, A. Felix, L. Moroder and C. Toniolo), Georg Thieme Verlag, Stuttgart.
8. Tawfik, D. (2002) Side Chain Selective Chemical Modifications of Proteins, in *The Protein Protocols Handbook* (ed. J.M. Walker), Humana Press, Totowa, NJ.
9. Hermanson, G.T. (1996) *Bioconjugate Techniques*, Academic Press, San Diego.
10. Offord, R. (1991) Going beyond the code. *Protein Engineering*, **4**, 709–710.
11. Offord, R.E. (1990) Chemical Approaches to Protein Engineering, in *Protein Design and the Development of New Therapeutics and Vaccines* (eds J.B. Hook and G.D. Poste), Plenum Press, New York.
12. Tolbert, T.J. and Wong, C.-H. (2002) New Methods for Proteomic Research: Preparation of Proteins with N-Terminal Cysteines for Labeling and Conjugation. *Angewandte Chemie-International Edition in English*, **41**, 2171–2174.
13. Dong, S.-Y., Ma, H.-M., Duan, X.-J., Chen, X.-Q. and Li, J. (2005) Detection of Local Polarity of alpha-Lactalbumin by N-Terminal Specific Labeling with a New Tailor-made Fluorescent Probe. *Journal of Proteome Research*, **4**, 161–166.
14. Mamaev, S., Olejnik, J., Olejnik, E.K. and Rothschild, K.J. (2004) Cell-free N-terminal protein labeling using initiator suppressor tRNA. *Analytical Biochemistry*, **326**, 25–32.
15. Dixon, H.B.F. and Fields, R. (1972) Specific modification of NH_2-terminal residues by transamination, in Methods in Enzymology (eds C.H.W. Hirs and S.N. Timasheff), Academic Press.
16. Gilmore, J.M., Scheck, R.A., Esser-Kahn, A.P., Joshi, N.S. and Francis, M.B. (2006) N-Terminal Protein Modification through a Biomimetic Transamination Reaction. *Angewandte Chemie-International Edition in English*, **45**, 5307–5311.
17. Dixon, H.B.F. (1984) N-terminal modification of proteins – a review. *Journal of Protein Chemistry*, **3**, 99–108.
18. Scheck, R.A. and Francis, M.B. (2007) Regioselective Labeling of Antibodies through N-Terminal Transamination. *ACS Chemical Biology*, **2**, 247–251.
19. Lewinska, M., Seitz, C., Skerra, A. and Schmidtchen, F.P. (2004) A Novel Method for the N-Terminal Modification of Native Proteins. *Bioconjugate Chemistry*, **15**, 231–234.
20. Plaut, A.G. and Bachovchin, W.W. (1994) IgA-specific prolyl endopeptidases: serine type. *Methods in Enzymology*, **244**, 137–151.
21. Deutch, C.E. (1984) Aminoacyl-tRNA: protein transferases. *Methods in Enzymology*, **106**, 198–205.

22 Giglione, C., Boularot, A. and Meinnel, T. (2004) Protein N-terminal methionine excision. *Cellular and Molecular Life Sciences*, **61**, 1455–1474.

23 Kuno, A., Taki, M., Kaneko, S., Taira, K. and Hasegawa, T. (2003) Leucyl/Phenylalanyl (L/F)-tRNA-protein transferase-mediated N-terminal specific labelling of a protein *in vitro*. *Nucleic Acids Symposium Series (Oxford)*, **3**, 259–260.

24 Taki, M. and Sisido, M. (2007) Leucyl/phenylalanyl(L/F)-tRNA-protein transferase-mediated aminoacyl transfer of a nonnatural amino acid to the N-terminus of peptides and proteins and subsequent functionalization by bioorthogonal reactions. *Biopolymers*, **88**, 263–271.

25 Taki, M., Kuno, A., Matoba, S., Kobayashi, Y., Futami, J., Murakami, H., Suga, H., Taira, K. et al. (2006) Leucyl/Phenylalanyl-tRNA-Protein Transferase-Mediated Chemoenzymatic Coupling of N-Terminal Arg/Lys Units in Post-translationally Processed Proteins with Non-natural Amino Acids. *Chembiochem*, **7**, 1676–1679.

26 Hecht, S.M., Alford, B.L., Kuroda, Y. and Kitano, S. (1978) Chemical aminoacylation of tRNA's. *Journal of Biological Chemistry*, **253**, 4517–4520.

27 Ibba, M., Becker, H.D., Stathopoulos, C., Tumbula, D.L. and Soll, D. (2000) The Adaptor hypothesis revisited. *Trends in Biochemical Sciences*, **25**, 311–316.

28 Copley, S.D. (2003) Enzymes with extra talents: moonlighting functions and catalytic promiscuity. *Current Opinion in Chemical Biology*, **7**, 265–272.

29 Budisa, N., Minks, C., Alefelder, S., Wenger, W., Dong, F., Moroder, L. and Huber, R. (1999) Toward the experimental codon reassignment *in vivo*: protein building with an expanded amino acid repertoire. *FASEB Journal*, **13**, 41–51.

30 Hohsaka, T., Sato, K., Sisido, M., Takai, K. and Yokoyama, S. (1993) Adaptability of nonnatural aromatic amino acids to the active center of the *E. coli* ribosomal A site. *FEBS Letters*, **335**, 47–50.

31 Olejnik, J., Gite, S., Mamaev, S. and Rothschild, K.J. (2005) N-terminal labeling of proteins using initiator tRNA. *Methods*, **36**, 252–260.

32 Gite, S., Mamaev, S., Olejnik, J. and Rothschild, K. (2000) Ultrasensitive Fluorescence-Based Detection of Nascent Proteins in Gels. *Analytical Biochemistry*, **279**, 218–225.

33 Taki, M., Sawata, S.Y. and Taira, K. (2001) Specific N-terminal biotinylation of a protein *in vitro* by a chemically modified tRNAfmet can support the native activity of the translated protein. *Journal of Bioscience and Bioengineering*, **92**, 149–153.

34 Tsalkova, T., Kramer, G. and Hardesty, B. (1999) The effect of a hydrophobic N-terminal probe on translational pausing of chloramphenicol acetyl transferase and rhodanese. *Journal of Molecular Biology*, **286**, 71–81.

35 Taki, M., Sawata, S.Y. and Taira, K. (2001) A novel immobilization method of an active protein via *in vitro* N-terminal specific incorporation system of nonnatural amino acids. *Nucleic Acids Symposium Series (Oxford)*, **1**, 197–198.

36 Brock, S., Szkaradkiewicz, K. and Sprinzl, M. (1998) Initiation factors of protein biosynthesis in bacteria and their structural relationship to elongation and termination factors. *Molecular Microbiology*, **29**, 409–417.

37 Gualerzi, C.O. and Pon, C.L. (1990) Initiation of mRNA translation in prokaryotes. *Biochemistry*, **29**, 5881–5889.

38 McIntosh, B., Ramachandiran, V., Kramer, G. and Hardesty, B. (2000) Initiation of protein synthesis with fluorophore-Met-tRNAf and the involvement of IF-2. *Biochimie*, **82**, 167–174.

39 Forster, A.C., Cornish, V.W. and Blacklow, S.C. (2004) Pure translation display. *Analytical Biochemistry*, **333**, 358–364.

40 Varshney, U. and RajBhandary, U. (1990) Initiation of Protein Synthesis from a Termination Codon. *Proceedings of the National Academy of Sciences of the United States of America*, **87**, 1586–1590.

41 Anthony-Cahill, S.J. and Magliery, T.J. (2002) Expanding the Natural Repertoire of Protein Structure and Function. *Current Pharmaceutical Biotechnology*, **3**, 299–315.

42 Murgola, E.J. (1985) tRNA, Suppression and the Code. *Annual Review of Genetics*, **19**, 57–80.

43 Xie, J. and Schultz, P.G. (2006) A chemical toolkit for proteins – an expanded genetic code. *Nature Reviews Molecular Cell Biology*, **7**, 775–782.

44 Connor, R.E. and Tirrell, D.A. (2007) Non-Canonical Amino Acids in Protein Polymer Design. *Polymer Review*, **47**, 9–28.

45 Bae, J.H., Rubini, M., Jung, G., Wiegand, G., Seifert, M.H.J., Azim, M.K., Kim, J.-S., Zumbusch, A. et al. (2003) Expansion of the Genetic Code Enables Design of a Novel "Gold" Class of Green Fluorescent Proteins. *Journal of Molecular Biology*, **328**, 1071–1081.

46 Kolb, H.C. and Sharpless, K.B. (2003) The growing impact of click chemistry on drug discovery. *Drug Discovery Today*, **8**, 1128–1137.

47 Tornøe, C.W., Christensen, C. and Meldal, M. (2002) Peptidotriazoles on Solid Phase: [1,2,3]-Triazoles by Regiospecific Copper(I)-Catalyzed 1,3-Dipolar Cycloadditions of Terminal Alkynes to Azides. *Journal of Organic Chemistry*, **67**, 3057–3064.

48 Rostovtsev, V.V., Green, L.G., Fokin, V.V. and Sharpless, K.B. (2002) A Stepwise Huisgen Cycloaddition Process: Copper(I)-Catalyzed Regioselective "Ligation" of Azides and Terminal Alkynes. *Angewandte Chemie-International Edition in English*, **41**, 2596–2599.

49 Beatty, K.E., Liu, J.C., Xie, F., Dieterich, D.C., Schuman, E.M., Wang, Q. and Tirrell, D.A. (2006) Fluorescence visualization of newly synthesized proteins in mammalian cells. *Angewandte Chemie-International Edition in English*, **45**, 7364–7367.

50 Beatty, K.E., Xie, F., Wang, Q. and Tirrell, D.A. (2005) Selective Dye-Labeling of Newly Synthesized Proteins in Bacterial Cells. *Journal of the American Chemical Society*, **127**, 14150–14151.

51 Deiters, A., Cropp, T.A., Mukherji, M., Chin, J.W., Anderson, J.C. and Schultz, P.G. (2003) Adding Amino Acids with Novel Reactivity to the Genetic Code of Saccharomyces Cerevisiae. *Journal of the American Chemical Society*, **125**, 11782–11783.

52 Lin, F.L., Hoyt, H.M., vanHalbeek, H., Bergman, R.G. and Bertozzi, C.R. (2005) Mechanistic Investigation of the Staudinger Ligation. *Journal of the American Chemical Society*, **127**, 2686–2695.

53 Saxon, E. and Bertozzi, C.R. (2000) Cell Surface Engineering by a Modified Staudinger Reaction. *Science*, **287**, 2007–2010.

54 Staudinger, H. and Meyer, J. (1919) Über neue organische Phosphorverbindungen III. Phosphinmethylenderivate und Phosphinimine. *Helvetica Chimica Acta*, **2**, 635–646.

55 Saxon, E., Armstrong, J.I. and Bertozzi, C.R. (2000) A "Traceless" Staudinger Ligation for the Chemoselective Synthesis of Amide Bonds. *Organic Letters*, **2**, 2141–2143.

56 Nilsson, B.L., Kiessling, L.L. and Raines, R.T. (2000) Staudinger Ligation: A Peptide from a Thioester and Azide. *Organic Letters*, **2**, 1939–1941.

57 Prescher, J.A., Dube, D.H. and Bertozzi, C.R. (2004) Chemical remodelling of cell surfaces in living animals. *Nature*, **430**, 873–877.

58 Thornton, J.M. and Sibanda, B.L. (1983) Amino and carboxy-terminal regions in globular proteins. *Journal of Molecular Biology*, **167**, 443–460.

59 Maden, B.E.H. (2003) Historical review: Peptidyl transfer, the Monro era. *Trends in Biochemical Sciences*, **28**, 619–624.

60 Nathans, D. (1964) Puromycin Inhibition of Protein Synthesis: Incorporation of Puromycin into Peptide Chains. *Proceedings of the National Academy of*

Sciences of the United States of America, **51**, 585–592.

61 Allen, D.W. and Zamecnik, P.C. (1962) The effect of puromycin on rabbit reticulocyte ribosomes. *Biochimica et Biophysica Acta*, **55**, 865–874.

62 Takeda, Y., Hayashi, S.-I., Nakagawa, H. and Suzuki, F. (1960) The effect of puromycin on ribonucleic acid and protein synthesis. *Journal of Biochemistry (Tokyo)*, **48**, 169–177.

63 Yarmolinsky, M.B. and Haba, G.L. (1959) Inhibition by puromycin of amino acid incorporation into protein. *Proceedings of the National Academy of Sciences of the United States of America*, **45**, 1721–1729.

64 Nathans, D. and Neidle, A. (1963) Structural Requirements for Puromycin Inhibition of Protein Synthesis. *Nature*, **197**, 1076–1077.

65 Miyamoto-Sato, E., Nemoto, N., Kobayashi, K. and Yanagawa, H. (2000) Specific bonding of puromycin to full-length protein at the C-terminus. *Nucleic Acids Research*, **28**, 1176–1182.

66 Nemoto, N., Miyamoto-Sato, E. and Yanagawa, H. (1999) Fluorescence labeling of the C-terminus of proteins with a puromycin analogue in cell-free translation systems. *FEBS Letters*, **462**, 43–46.

67 Doi, N., Takashima, H., Kinjo, M., Sakata, K., Kawahashi, Y., Oishi, Y., Oyama, R., Miyamoto-Sato, E. *et al.* (2002) Novel Fluorescence Labeling and High-Throughput Assay Technologies for *In Vitro* Analysis of Protein Interactions. *Genome Research*, **12**, 487–492.

68 Kobayashi, T., Shiratori, M., Nakano, H., Eguchi, C., Shirai, M., Naka, D. and Shibui, T. (2007) Short peptide tags increase the yield of C-terminally labeled protein. *Biotechnology Letters*, **29**, 1065–1073.

69 Oyama, R., Takashima, H., Yonezawa, M., Doi, N., Miyamoto-Sato, E., Kinjo, M. and Yanagawa, H. (2006) Protein-protein interaction analysis by C-terminally specific fluorescence labeling and fluorescence cross-correlation spectroscopy. *Nucleic Acids Research*, **34** (e), 102.

70 Yamaguchi, J., Nemoto, N., Sasaki, T., Tokumasu, A., Mimori-Kiyosue, Y., Yagi, T. and Funatsu, T. (2001) Rapid functional analysis of protein-protein interactions by fluorescent C-terminal labeling and single-molecule imaging. *FEBS Letters*, **502**, 79–83.

71 Kawahashi, Y., Doi, N., Takashima, H., Tsuda, C., Oishi, Y., Oyama, R., Yonezawa, M., Miyamoto-Sato, E. *et al.* (2003) *In vitro* protein microarrays for detecting protein-protein interactions: application of a new method for fluorescence labeling of proteins. *Proteomics*, **3**, 1236–1243.

72 Humenik, M., Huang, Y., Wang, Y. and Sprinzl, M. (2007) C-Terminal Incorporation of Bio-Orthogonal Azide Groups into a Protein and Preparation of Protein-Oligodeoxynucleotide Conjugates by Cu^{I}-Catalyzed Cycloaddition. *Chembiochem*, **8**, 1103–1106.

73 Starck, S.R., Green, H.M., Alberola-Ila, J. and Roberts, R.W. (2004) A General Approach to Detect Protein Expression *In Vivo* Using Fluorescent Puromycin Conjugates. *Chemistry & Biology*, **11**, 999–1008.

9
Using the Bacteriophage MS2 Coat Protein–RNA Binding Interaction to Visualize RNA in Living Cells

Jeffrey A. Chao, Kevin Czaplinski, and Robert H. Singer

9.1
Introduction

The development of naturally fluorescent proteins (FPs) with various photophysical properties whose emission wavelengths span most of the visible spectrum has allowed biologists to characterize dynamic processes in real time in living cells. These genetically encoded FPs obviate the need for invasive techniques to introduce the fluorescent label, such as microinjection, that may alter celluar physiology. In order to take advantage of the expanding array of FPs for studying the metabolism of RNA in real time, a variety of strategies have been employed.

Fluorescent protein technology has also been used in the study of nucleic acids in real time. One straightforward approach for characterizing RNA is to generate a chimeric fusion between a known RNA-binding protein and a FP. Two such examples, FMRP-GFP and ZBP1-GFP have been used to track the movement of large granules that contain many RNAs in PC12 cells and chicken embryonic fibroblasts [1, 2]. A second strategy has recently been developed that allows direct labeling of endogenous RNA. This strategy used two Pumilio homology domains (PUF domains) that have been engineered to recognize distinct octamer RNA sequences that are adjacent to each other in the ND6 mtRNA, a mitochondrial RNA encoding NADH dehydrogenase subunit 6 [3]. The PUF domains were fused to either N- or C-terminal fragments of GFP (bimolecular fluorescence complementation; BiFC), which independently are not fluorescent. Upon RNA-binding to the adjacent sites, the PUF-GFP fragments are in close enough proximity to allow association of the GFP fragments and maturation of the fluorophore. BiFC has also been used to show RNA binding site dependent interactions between RNA binding proteins [4].

A limitation, however, to both of these techniques is that they are unable to detect RNAs with single transcript resolution. To achieve this level of sensitivity,

Probes and Tags to Study Biomolecular Function. Lawrence W. Miller (Ed.)
Copyright © 2008 WILEY-VCH Verlag GmbH & Co. KGaA, Weinheim
ISBN: 978-3-527-31566-6

methodologies have been developed that use RNA reporters that contain multiple binding sites for a chimeric RNA-binding protein fused to a FP. The specific interaction between the bacteriophage MS2 coat protein (MS2 CP) and its cognate RNA hairpin binding site (MBS) has been exploited in the yeast three-hybrid system, affinity purification, tethering, and fluorescent labeling of RNAs in living cells [5–8]. The MS2 CP (129 amino acids, MW $\approx$ 14 kDa) forms a dimer that assembles into the bacteriophage capsid and also regulates transcription of the bacteriophage replicase by binding a small RNA hairpin (19 nucleotides) that contains the replicase start codon. The MS2 CP binds its wild type RNA target with high affinity (low nanomolar dissociation constant) and both protein and RNA mutants have been identified that increase affinity of this interaction (Val29Ile and Uri5Cyt) [9–11]. This exquisitely tight and specific interaction allows stable fluorescent labeling of the reporter RNA construct during live cell imaging.

Alternative systems for labeling RNAs with FPs have also been developed. One system utilizes the interaction between the human splicing factor U1A and the polyadenylation inhibition element located in the 3' UTR of its pre-mRNA and another employs eIF4a and a high affinity RNA aptamer target [12, 13]. These systems have been applied to imaging of RNAs in yeast and bacteria, however, they are not suitable for studies in higher eukaryotes because they will cross-react and compete with endogenous RNA–protein complexes. Since the MS2 RNA reporter system is derived from an ssRNA bacteriophage it is orthogonal to endogenous RNA–protein complexes in both prokaryotes and eukaryotes and has been used in a variety of applications and experimental systems.

The MS2 reporter system has been used successfully to study many aspects of gene expression, ranging from transcription to RNA localization. In *Escherichia coli*, the kinetics of gene expression were measured by quantifying mRNA levels in single bacteria cells [14]. The pulsating transcriptional behavior of a developmental gene, *dscA*, in *Dictyostelium* has also been characterized [15]. In a human osteosarcoma cell line (U2OS), the MS2 system has been used to study transcriptional activation [16], the diffusion of mRNAs in the nucleus [17], and the kinetics of PolII transcription have been measured by quantifying the fluorescent recovery after photobleaching (FRAP) of an MBS-containing reporter mRNA [18]. The movement and localization of mRNA has been characterized in yeast [7], plants [19], *Drosophila* [20], primary rodent neurons [21, 22], and COS cells [23] (Figure 9.1).

In this chapter, we will outline how to use the MS2 system for imaging RNAs with single transcript resolution in mammalian cell lines. The success of detection and analysis of RNAs in living cells is largely dependent upon the RNA and protein reporter constructs that are used. The first section of this chapter will describe strategies for generating and expressing MBS containing RNAs and chimeric MS2 CP fusions to fluorescent proteins. This will be followed by a summary of a live cell imaging microscopy platform and a general protocol for data collection. We conclude the chapter by discussing methods for data analysis.

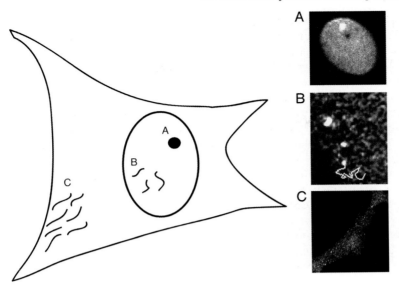

Figure 9.1 Imaging RNAs in living cells. Real-time analysis of RNAs in living cells allows gene expression to be quantitatively measured at many stages throughout the lifespan of an RNA. (A) Transcription kinetics measured by fluorescence recovery after photobleaching (Reprinted from [18]). (B) Quantification of RNA movements in nucleoplasm (Reprinted from [17]). Characterization of β-actin mRNA in cytoplasm (Reprinted from [23]).

9.2
Construction of an MBS-Containing Reporter RNA

In practice, precise control of MS2 CP-FP expression is difficult and unbound molecules of MS2 CP-FP will be present in the sample. Consequently, the intensity of the fluorescent signal of reporter RNA with a single MBS will be indistinguishable from a free MS2 CP-FP probe. As a result, to detect single RNA molecules above the background of free MS2 CP-FP it is necessary to multimerize the MBS so that individual RNAs contain many copies of the MS2 CP-FP. Furthermore, it has been shown that the MS2 CP binds cooperatively to adjacent hairpins, thereby increasing the occupancy of the binding sites [24]. It has been empirically determined that 24 copies of the MBS (binding a maximum of 48 MS2 CP-FP) provides a sufficient signal-to-noise ratio (SNR) for most applications. Plasmids containing MBS cassettes with 6, 12, and 24 copies of the binding site (pSL-MS2-6X, -12X, -24X) can be obtained from the Singer lab (http://singerlab.org/requests/). In initial studies 6x and 12x MBS sites did not yield visible particles, however these MBS cassettes can serve as a starting point for the construction of reporter RNA constructs [23].

The 24X MBS cassette from pSL-24X is 1.4 kb and the repetition of the MS2 stem-loops presents some challenges for generating reporter RNAs. Propagation of plasmids that contain the 24X MBS cassette has a propensity to deplete copies of the stem-loop in bacteria via homologous recombination. Bacterial strains that have been engineered for lower recombination rates, like Stbl2 (Invitrogen Cat#10268019), increase retention of the 24X MBS cassette, however, we have found that it is often still necessary to screen several colonies to identify clones that contain the full-length cassette. The 24X MBS cassette can be excised from the vector backbone by restriction digest with BamHI and BglII and ligated into a reporter construct cut with BamHI, BglII or BclI, which have compatible cohesive ends. RNA reporter constructs that necessitate the use of other restriction enzymes can be accommodated by PCR amplification of the MBS cassette, which requires primers outside the repeats, or blunt-ended cloning.

The position of the 24X MBS cassette within the reporter RNA is important to minimize perturbation of the transcript caused by insertion of the exogenous sequence. Inserting the cassette into an mRNA within regions responsible for post-transcriptional control will interfere with the regulation of the mRNA reporter construct. Initiation of nonsense mediated decay (NMD), disruption of cis-acting elements that mediate mRNA processing or post-transcriptional regulation and changes in mRNA stability are potential issues that should be avoided when designing a reporter RNA. In several examples, placement of the 24X MBS cassette either between the stop codon of the ORF and the 3′UTR or proximal to the polyadenylation site in the 3′UTR has been found to result in functional mRNAs [7, 17, 23]. To help choose an insertion site, sequence alignments across related species can be used as a guideline. Locations within the 3′UTR that are less evolutionary conserved are good candidates for insertion sites because they may be less likely to disrupt elements important for regulation. The effect of inserting the MBS cassette may be context dependent and should be evaluated for each reporter RNA.

Once the reporter RNA construct has been generated, it must then be introduced into cells. The RNA reporter can be either transiently or stably expressed, depending upon the requirements of the experimental system. Transient expression can be achieved with calcium phosphate co-precipitation, standard lipid-based reagents or electroporation. Stable expression of the reporter RNA can be obtained by selection if the construct contains a resistance marker. Integration of the reporter RNA occurs at a random position within the genome and may result in a tandem array containing multiple copies of the construct. For this reason, it is necessary to select and characterize individual clones. Recent advances in gene expression technology, which take advantage of retroviral and lentiviral vectors, facilitate the generation of stable cell lines with single integrations and may be appropriate for some experimental systems.

9.3
Construction of an MS2 CP-FP Chimera

In order to visualize the reporter RNA in living cells, it is necessary to co-introduce an MS2 CP-FP chimera. Due to the demanding imaging requirements necessary for single molecule detection, the brightness and photostability of the FP are important

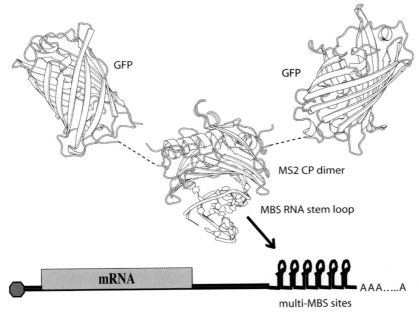

Figure 9.2 The bipartite MS2-CP MBS RNA tag. A structural diagram of the MS2-CP dimer bound to a single MBS stem loop is shown based on the solved co-crystal structure. GFP is fused to the C-terminus of the MS2-CP. An MBS tagged substrate engineered with multiple MBS sites encoded in the 3′UTR is shown, the arrow reflects that each MBS stem loop added specifically recruits one MBS dimer, such that the number of GFPs targeted to the substrate is equal to twice the number of MBS sites added.

criteria that must be evaluated before selecting an FP. Both eGFP and eYFP have been used successfully and therefore have sufficient brightness to allow for detection of single mRNAs using the MS2 system (Figure 9.2). There are, however, an increasing number of options available for the FP portion of the chimera including TagRFP, a bright monomeric red fluorescent protein, that may be suitable for some applications [25].

The expression level of the MS2 CP-FP fusion is a critical parameter for the success of single molecule detection of RNAs because both over- and under-expression will reduce the SNR ratio of reporter RNA particles. Different approaches can be used to help achieve appropriate levels of the MS2 CP-FP by modifying the overall expression of the fusion protein or by targeting the fusion protein to different sub-cellular locations within the cell. Expression of the MS2 CP-FP can be controlled by constitutive promoters of varying strength and expression vectors with the promoters from cytomegalovirus, the large subunit of RNA polymerase II and ribosomal protein L30 are available from the Singer lab. Inducible and tissue-specific promoters can also be used to regulate expression if more stringent control is required. In conjunction with varying the overall expression levels, it is also possible to deplete unbound MS2 CP-FP from the cytoplasm by including a nuclear localization signal

(NLS) in the construct. This strategy is particularly effective for imaging RNA movements in the cytoplasm, but can also be used with low expression levels to visualize nuclear events. Finally, cells stably expressing MS2 CP-FP can be partitioned by fluorescence activated cell sorting (FACS) for desirable expression levels.

9.4
Co-introduction of MS2 Reporter RNA and MS2 CP-FP

Both elements of the MS2 system must be expressed within the same cells in order to visualize the mRNA in real time. Multiple plasmids can easily be co-expressed by transient transfection. Using this approach one must be cautious about high levels of expression of MS2-CP that typically result from transient transfection and approaches described above may be used to help obtain appropriate expression. Moreover, a second marker for co-transfection of the reporter RNA is required. One strategy that can overcome the limitation of very high expression during transient transfection is to transfect using electroporation followed immediately by plating cells on a substrate-coated dish to facilitate attachment and spreading [17]. The kinetics of expression are more consistent using electroporation than other transfection methods and electroporated cells are imaged at time points where expression begins to be visible. Under these conditions, many cells reach visible levels of expression concomitantly at times when few have a high expression of MS2 CP-FP (see transfection protocol). The limitation of this approach is that it is only compatible with cell types amenable to electroporation.

A more reliable method for generating cells for imaging mRNAs is to create stable lines. Cell lines can be generated that stably express either the reporter mRNA, the MS2 CP-FP or both components. Stably transfected cells lines can be created in the classical way by transient transfection, followed by selection with a resistance marker co-expressed with the construct of interest. Individual clones can then be selected for desirable expression levels. Once a stable line is created with one component of the system the complementary construct can be introduced by single transfection into that cell line. For example, a stable line expressing MS2 CP-FP can be subsequently transiently transfected with a single plasmid expressing the MBS tagged mRNA. In our experience, creating MS2 CP-FP stable expression lines makes data acquisition more robust and reproducible.

9.5
Microscopy Platform for Single Molecule Detection of RNAs in Living Cells

In order to capture the movements of mRNA in real time the imaging system should be optimized for both single molecule sensitivity and high-speed acquisition. If the goal of the experiment is to be able to track mRNA movement in real time, the temporal and spatial resolution that needs to be achieved must be carefully considered. For instance, active mRNA transport has been reported to move at as fast as $4\,\mu m\,s^{-1}$ through the processes of primary neurons. With a pixel size of 106.7 nm (U Apochromat 150×1.45 NA objective (Olympus) and Cascade II 512 EMCCD camera (Photometrics)), during a 100 ms exposure, a particle moving at this velocity will move approximately four pixels. To ensure accurate tracking of a particle in three dimensions, images with sufficient SNR must be acquired at a minimum rate of twice the frequency (Nyquist sampling) of the movement, which becomes challenging if multiple z-planes are required. In the following section, the core components of an imaging system optimized for live cell imaging will be described. This system can be further tailored to meet specific experimental requirements.

9.5.1
Components of the Imaging System

Microscope stand and stage: Inverted microscope (IX81, Olympus) equipped with a motorized stage and fast axial position controller (MS-2000-XYLE-PZ, Applied Scientific Instruments Inc.) The motorized stage allows multi-field imaging while the axial controller enables rapid collection of z-stacks. For many experimental applications, an auto-focus mechanism will facilitate the acquisition of time-lapse data sets (ZDC, Olympus).

Camera: Electron multiplying charge coupled device (EMCCD) cameras (Cascade II:1024, Photometrics) provide both high sensitivity and rapid image acquisition and are required for the low light levels inherent in single molecule imaging.

Light source: Illumination of the specimen can be accomplished using either mercury or xenon lamps or laser light sources. In order to reduce the effects of phototoxicity and photobleaching, it may be necessary to attenuate the intensity of the light using neutral density filters. Another consideration for light source selection is the ability to switch rapidly between wavelengths for multicolor imaging applications.

Objective: A high numerical aperture (NA) objective (U Apochromat 150×1.45 NA, Olympus) allows for efficient collection of light from the sample (brightness) and for resolution of fine cellular details. Objectives with other properties including compatibility with total internal reflection fluorescence microscopy (TIRFM) and the ability to transmit UV light for photoactivation and uncaging experiments may also be considered.

Environmental Control: To maintain cellular physiology and viability during imaging it is necessary to maintain the temperature of the sample chamber at 37 °C (Delta T temperature controller, Bioptechs Inc). Additionally, CO_2 and humidity levels can also be regulated during imaging if cells are to be maintained for extended periods of time on the microscope and many environmental chambers that control atmosphere and temperature are available for any microscopy set-up.

9.6
Protocols for Co-expressing MS2 CP-FP- and MBS-Containing Plasmids for Live Cell Imaging

Using the following protocol we have co-transfected U2OS or COS cells, each of which have well-spread morphology that suits live imaging of cellular events very well. This protocol is diagrammed in Figure 9.3. Using this method it is advantageous to image cells at time points post-electroporation where MCP-FP is clearly visible, but not yet overwhelming [16, 17].

- 10 cm dish of U2OS cells are grown to 70% confluence (medium: DMEM/10% FBS)
- remove media and detach cells with trypsinization
- quench trypsin when cells have detached
- pellet cells by centrifugation
- remove supernatant and resuspend cells in 600 µl of medium

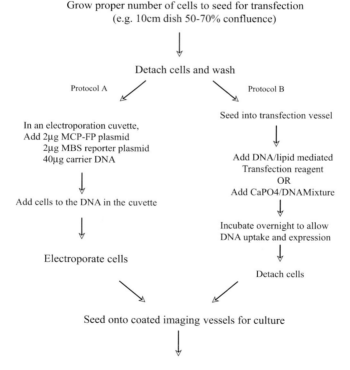

Figure 9.3 A generalized flowchart for two alternative approaches to the co-transfection of MCP-FP and MBS reporter plasmid DNA.

- separately, add 2 μg of pMS2-GFP-NLS and 2 μg MBS containing plasmid along with 40 μg sheared salmon sperm DNA to a 4 mm gap electroporation cuvette
- add 200 μl of cell suspension to the DNA containing cuvette
- electroporate (170 V, 950 μF)
- transfer cells to 2 ml of fresh medium
- centrifuge cells and remove supernatant
- resuspend pellet in DMEM with HEPES lacking Phenol red and riboflavin/10% FBS
- these electroporated cells can now be seeded in any coated culture chamber that is suitable for high optical resolution imaging (we typically use the Bioptechs FCS or DeltaT open dish system). For culture dish coating we used Cell-Tak (BD Biosciences), but any other commonly used coating that allows quick attachment and spreading of cells should produce similar results. Maintenance of cells in DMEM without Phenol Red/and riboflavin helps reduce autofluorescent background that comes from the medium. Leibovitz's L-15 medium with FBS is an alternative medium without fluorescent background
- incubate cells for several hours and begin imaging session as soon as clearly visible levels of MCP-GFP are achieved.

Similarly, co-transfection can be achieved by calcium phosphate transfection or lipid-mediated transfection in an appropriate culture vessel and standard protocols for these common procedures are widely available. Since there is a very wide variety of lipid-mediated transfection reagents available, following the manufacturers guidelines and testing optimal DNA:transfection reagent ratios is advisable for each construct as well as each cell line. Transfection is followed by trypsinization 24 h post-transfection and re-plating on Cell-tak coated culture chambers suitable for high resolution imaging [23]. Imaging is performed as soon as the cells have attached. In this protocol cells can be imaged as soon as they have attached to the substrate and spread to a morphology that accommodates imaging. One study using primary neurons has used lisitc co-delivery of the plasmids using a gene gun [22]. In this protocol the plasmid DNAs are mixed with microscopic gold particles under conditions that allow the DNA to bind to gold particles. The particles are then shot onto the culture under conditions that allow particles to enter the cells, delivering the DNA across the plasma membrane where the transfected cells remove the DNA from the beads and express the encoded construct. Any other method for co-transfection plasmids should therefore be suitable for this procedure.

9.7
Image Analysis and Quantification of mRNA Molecules

Before undertaking a live cell experiment, it is advisable to characterize the behavior of the MBS-containing reporter mRNA and the MS2 CP-FP in fixed cells. The spatial distribution and number of reporter RNAs can be quantified by fluorescent *in situ* hybridization (FISH) using DNA oligonucleotide probes conjugated to fluorescent dyes that are anti-sense to the reporter RNA [23, 26, 27]. Additionally, a FISH probe against the MBS cassette provides high quality images with excellent SNR because many probes with multiple fluorescent dyes will hybridize to a single RNA. Once the experimental system has been adequately characterized in fixed cells, live cell imaging experiments can commence.

Quantification of FP Fluorescence: Single molecule measurements of RNAs labeled with MS2 CP-FP fusions can be performed in live cells by comparing the total fluorescence intensity (TFI) of a reporter RNA to a calibration curve of the purified FP. Serial dilutions of the purified FP can be used to determine the TFI of specific concentrations of the FP per voxel, that can then be converted to the fluorescence intensity of a single FP using a linear regression analysis. A detailed description of this method can be obtained from Femino *et al.* [27]. It is vital that an identical optical configuration and acquisition parameters, including camera gain and binning, are used to image the calibration curve samples and the experimental specimen to ensure accurate comparison. Once the calibration curve is generated, the number of reporter mRNAs in the cell can then be calculated. Using this method, it was determined that an average of 33 MS2 CP-GFP molecules bound a reporter RNA containing 24 copies of the MBS [23].

Tracking single mRNA particles: The spatio-temporal tracking of RNA particles within living cells allows their dynamic behavior to be directly observed and provides insights into the molecular mechanisms underlying their motions. A more detailed description of the algorithms used to identify and track particles can be found in Levi and Gratton [28]. Several particle tracking packages are available as plug-ins for ImageJ (http://rsb.info.nih.gov/ij) that allow automatic tracking of fluorescent particles within time lapse movies and calculation of their trajectories and velocities [29, 30]. Using single particle tracking, RNA molecule movements can be classified as either directed, corralled, diffused or stationary, and the frequency of these movements can be altered by sequences within the mRNA.

References

1 De Diego Otero, Y., Severijnen, L.A. *et al.* (2002) Transport of fragile X mental retardation protein via granules in neurites of PC12 cells. *Molecular and Cellular Biology*, **22** (23), 8332–8341.

2 Oleynikov, Y. and Singer, R.H. (2003) Real-time visualization of ZBP1 association with beta-actin mRNA during transcription and localization. *Current Biology*, **13** (3), 199–207.

3 Ozawa, T., Natori, Y. *et al.* (2007) Imaging dynamics of endogenous mitochondrial RNA in single living cells. *Nature Methods*, **4** (5), 413–419.

4 Rackham, O. and Brown, C.M. (2004) Visualization of RNA-protein interactions in living cells: FMRP and IMP1 interact on mRNAs. *EMBO Journal*, **23** (16), 3346–3355.

5 Bardwell, V.J. and Wickens, M. (1990) Purification of RNA and RNA-protein complexes by an R17 coat protein affinity method. *Nucleic Acids Research*, **18** (22), 6587–6594.

6. SenGupta, D.J., Zhang, B. et al. (1996) A three-hybrid system to detect RNA-protein interactions in vivo. *Proceedings of the National Academy of Sciences of the United States of America*, **93** (16), 8496–8501.

7. Bertrand, E., Chartrand, P. et al. (1998) Localization of ASH1 mRNA particles in living yeast. *Molecules and Cells*, **2** (4), 437–445.

8. Lykke-Andersen, J., Shu, M.D. et al. (2000) Human Upf proteins target an mRNA for nonsense-mediated decay when bound downstream of a termination codon. *Cell*, **103** (7), 1121–1131.

9. Carey, J., Cameron, V. et al. (1983) Sequence-specific interaction of R17 coat protein with its ribonucleic acid binding site. *Biochemistry*, **22** (11), 2601–2610.

10. Lowary, P.T. and Uhlenbeck, O.C. (1987) An RNA mutation that increases the affinity of an RNA-protein interaction. *Nucleic Acids Research*, **15** (24), 10483–10493.

11. Lim, F. and Peabody, D.S. (1994) Mutations that increase the affinity of a translational repressor for RNA. *Nucleic Acids Research*, **22** (18), 3748–3752.

12. Brodsky, A.S. and Silver, P.A. (2002) Identifying proteins that affect mRNA localization in living cells. *Methods*, **26** (2), 151–155.

13. Valencia-Burton, M., McCullough, R.M. et al. (2007) RNA visualization in live bacterial cells using fluorescent protein complementation. *Nature Methods*, **4** (5), 421–427.

14. Golding, I., Paulsson, J. et al. (2005) Real-time kinetics of gene activity in individual bacteria. *Cell*, **123** (6), 1025–1036.

15. Chubb, J.R., Trcek, T. et al. (2006) Transcriptional pulsing of a developmental gene. *Current Biology*, **16** (10), 1018–1025.

16. Janicki, S.M., Tsukamoto, T. et al. (2004) From silencing to gene expression: real-time analysis in single cells. *Cell*, **116** (5), 683–698.

17. Shav-Tal, Y., Darzacq, X. et al. (2004) Dynamics of single mRNPs in nuclei of living cells. *Science*, **304** (5678), 1797–1800.

18. Darzacq, X., Shav-Tal, Y. et al. (2007) In vivo dynamics of RNA polymerase II transcription. *Nature Structural & Molecular Biology*, **14** (9), 796–806.

19. Zhang, F. and Simon, A.E. (2003) A novel procedure for the localization of viral RNAs in protoplasts and whole plants. *Plant Journal*, **35** (5), 665–673.

20. Forrest, K.M. and Gavis, E.R. (2003) Live imaging of endogenous RNA reveals a diffusion and entrapment mechanism for nanos mRNA localization in Drosophila. *Current Biology*, **13** (14), 1159–1168.

21. Bi, J., Tsai, N.P. et al. (2006) Axonal mRNA transport and localized translational regulation of kappa-opioid receptor in primary neurons of dorsal root ganglia. *Proceedings of the National Academy of Sciences of the United States of America*, **103** (52), 19919–19924.

22. Dynes, J.L. and Steward, O. (2007) Dynamics of bidirectional transport of Arc mRNA in neuronal dendrites. *Journal of Comparative Neurology*, **500** (3), 433–447.

23. Fusco, D., Accornero, N. et al. (2003) Single mRNA molecules demonstrate probabilistic movement in living mammalian cells. *Current Biology*, **13** (2), 161–167.

24. Witherell, G.W., Wu, H.N. et al. (1990) Cooperative binding of R17 coat protein to RNA. *Biochemistry*, **29** (50), 11051–11057.

25. Merzlyak, E.M., Goedhart, J. et al. (2007) Bright monomeric red fluorescent protein with an extended fluorescence lifetime. *Nature Methods*, **4** (7), 555–557.

26. Femino, A.M., Fay, F.S. et al. (1998) Visualization of single RNA transcripts in situ. *Science*, **280** (5363), 585–590.

27. Femino, A.M., Fogarty, K. et al. (2003) Visualization of single molecules of mRNA in situ. *Methods in Enzymology*, **361**, 245–304.

28. Levi, V. and Gratton, E. (2007) Exploring dynamics in living cells by tracking single particles. *Cell Biochemistry and Biophysics*, **48** (1), 1–15.

29 Sage, D., Neumann, F.R. et al. (2005) Automatic tracking of individual fluorescence particles: application to the study of chromosome dynamics. *IEEE Transactions on Image Processing*, **14** (9), 1372–1383.

30 Sbalzarini, I.F. and Koumoutsakos, P. (2005) Feature point tracking and trajectory analysis for video imaging in cell biology. *Journal of Structural Biology*, **151** (2), 182–195.

Index

a
A1 tag   131f.
acyl carrier protein (ACP)   123ff.
– synthase AcpS   121ff.
aerolysin   66
– GPI-anchored protein   66
O^6-alkylguanine-DNA alkyltransferase (AGT)   90, 121ff.
– tag   89ff.
amino acid
– non-canonical   153ff.
aminoacyl-tRNA synthase (AARS)   147
AOT   12f.

b
bacteriophage   163ff.
BCθ (biotinylated Cθ-toxin)   6, 64
22-(p-benzoylphenoxy)-23,24-bisnorcholan-5-en-3b-ol (FCBP) photoactivatable sterol   20
O^6-benzylguanine (BG)   90
– functionalized surface   104
biarsenical-tetracysteine system   74ff.
– protocol   82ff.
biomimetic transamination   140ff.
bioorthogonal chemical transformation   139ff.
biotin carboxyl carrier protein (BCCP)   122
biotin ligase BirA   122
BODIPY-cholesterol   4ff., 14ff.
BODIPY PI   38
BODIPY TMP   113

c
carrier protein fusion   127
caveolae   2
cell
– labeling in living cell   102, 109ff.
– single molecule detection of RNA in living cell   168
cell surface protein labeling   100
chemical inducer of dimerization (CID)   102
chemical transformation   140ff.
– expanded genetic code   145
– protein C-terminus   155
cholera toxin (CT)   54
– intracellular trafficking   56
cholera toxin B subunit (CTB)   54ff.
– protocol   57
cholesterol   3
– dansyl, see dansyl-cholesterol
– distribution and dynamics   53ff.
– distribution in membrane   7
– NBD   4ff.
– photoactivatable   4
– source   3
– structure   7
– trafficking   1ff.
cholesterol-binding toxin   63
cholesterol-dependent cytolysin (CDC)   63
cholesterol-rich domain   64f.
click chemistry   153ff.
coenzyme A (CoA)   123ff.
– fluorophore   123ff.
confocal laser scanning microscopy (CLSM)
– sterol probe   6
covalent immobilization
– SNAP-tag   104
N^9-cyclopentyl-O^6-(4-bromothenyl)guanine (CG)   94

d
dansyl-cholesterol (6-dansyl-cholestanol, DChol)   4ff., 13f.
DHE (dehydroergosterol)   4ff., 17ff.

– methyl-β-cyclodextrin (MβCD)   18f.
dihydrofolate reductase (DHFR) fusion protein   109ff.
dimyristolphosphatidylcholine (DMPC)   9
dioleoylphosphatidylcholine   59
dipalmitoylphosphatidylcholine (DPPC)   11, 59
DOPS (1,2-dioleoyl-sn-glycero-3-[phospho-L-serine])   5
DRM, see membrane
dye
– non-permeable   100

e

endocytic recycling compartment (ERC)   18
enzymatic modification of N-terminus   143
epidermal growth factor (EGF)   43
expression vector   112

f

fatty acid synthase (FAS)   123
FCBP photoactivatable sterol cross-linker   4
FlAsH (Fluorescein Arsenical Hairpin Binder)   74ff.
– protein labeling   73ff.
– protocol   83
fluorescence correlation spectroscopy (FCS)   55
fluorescence loss in photobleaching (FLIP)   18
fluorescence recovery after photobleaching (FRAP)   18, 55, 164
fluorescence resonance energy transfer (FRET)   55, 128f.
– conformational change   101
– protein–protein interaction   101
fluorescent biarsenical ligand   75
fluorescent sterol   1ff.
– analog   4
– labeling methology   3ff.
– methyl-β-cyclodextrin (FS-MβCD)   5
– source   3
fluorophore
– alkyne   153
– azido   153

g

genetic code   140ff.
– expansion   145f.
GM1 (Galβ1,3GalNAcβ1,4(NeuAcα2,3)Galβ1,4GlcCer)   54ff.
– fluorescent-labeled   58
GPI-anchored protein   66
green fluorescent protein (GFP)   41ff., 163

h

high density lipoprotein (HDL)   5
Huisgen [3+2] cycloaddition   153ff.

i

image analysis   171f.
– mRNA   171f.
imaging
– phosphoinositide probe   40
immunoglobulin A (IgA) protease   143
inositol phosphate   48
ionomycin   45

l

L/F-tRNA protein transferase   143ff.
large unilamellar vesicle (LUV)   5, 18
lipid binding protein   35ff.
lipid binding toxin   53ff.
lipid raft   2, 55
live cell imaging   170
– protocol   170
lysenin   59ff.
– non-toxic (NT-Lys)   61ff.
– protocol   62

m

MBS (MS2CP binding site)   164ff.
– containing reporter RNA   165ff.
– protocol   170
MCθ (methylated Cθ-toxin)   64
membrane
– detergent-resistant (DRM)   2, 55ff.
microscopy in living cell   168f.
molecular imaging   121, 169
MS2 coat protein (MS2CP)   164ff.
– protocol   170
multiphoton laser scanning microscopy (MPLSM)
– sterol probe   6

n

NBD cholesterol   4ff.
22-NBD cholesterol (22-(N-(7-nitrobenz-2-oxa-1,3-diazol-4-yl)amino)-23,24-bisnor-5-cholen-3β-ol)   9ff.
25-NBD cholesterol (25-(N-[(7-nitrobenz-2-oxa-1,3-diazol-4-yl)-methyl]amino)-27-norcholesterol)   11ff.
NBD PtdIns   38
nonribosomal peptide synthetase (NRPS)   123

o

oocyte   43ff.

p

peptidyl carrier protein (PCP)   123ff.
perfringolysin O (PFO, θ toxin)   63ff.
– protocol   65
PH (pleckstrin homology) domain
– GFP-tag   41f.
phosphatidylinositol (PtdIns)   35f.
– detection of synthesis   41ff.
– PtdIns(3,4,5)P$_3$ 35ff.
– PtdIns(4,5)P$_2$ 35ff.
phosphoinositide (PI)   35ff.
– fluorescent derivative   38
– fluorescent PI-binding domain   39
– imaging   40
– monitoring distribution and dynamics   37
– probe   40ff.
– signalling   35
phosphoinositide 3-kinase (PI3K)   42f.
phospholipase C (PLC) family   35f.
phosphopantetheinyl transferase (PPTase)   122
– protein translational modification   123
photobleaching   99
poly(oligo(ethylene glycol)methacrylate) (POEGMA)   105
polyketide synthase (PKS)   123
POPC (1-palmitoyl-2-oleoyl-sn-glycero-3-phosphocholine)   5
– large unilamellar vesicle (LUV)   9ff.
posttranslational modification (PTM)   121ff.
O^6-propargylguanine (PG)   94
protein
– bioorthogonal chemical transformation   139ff.
– fluorescent (FP), *see also* green fluorescent protein   163ff.
– internal modification   149f.
protein dimerization   102
protein labeling   89ff., 114
– carrier protein fusion   127
– cell-free   147ff.
– cell surface   100
– covalent   102
– FlAsH (Fluorescein Arsenical Hairpin Binder)   73ff.
– intracellular   101
– multicolor   101
– N-terminal   140ff.
– orthogonal   131f.
– phosphopantetheinyl transferase catalyzed   121ff.
– short peptide tag   131
– tRNA   147ff.

protein-protein interaction   101
PUF domain (Pumilio homology domain)   163
pulse chase labelling   100
puromycin   155ff.

r

ReAsH   76ff.
RNA
– binding protein   163
– cRNA   44
– initiator tRNA   147
– L/F-tRNA protein transferase   143
– MBS-containing reporter   164ff.
– mRNA   166ff.
– single molecule detection   167
– suppressor initiator tRNA   148

s

S6 tag   131f.
Schizosaccharomyces cerevisiae   79ff.
– biarsenical-tetracysteine system   82ff.
sensor   102
Sfp   122ff.
Sharpless-Meldal "click" chemistry   153
single-particle tracking (SPT)   55
single molecule detection   168
SNAP-tag   89ff., 101
– cell imaging   95f.
– covalent immobilization   104
– labeling   91ff.
– protocol   97
– protein function   102
sphingolipid
– distribution and dynamics   53ff.
sphingomyelin   59ff.
– probe   60
Staudinger–Bertozzi ligation   153
Staudinger ligation   153ff.
sterol
– fluorescent, *see* fluorescent sterol
sterol carrier protein-2 (SCP-2)   10f.
stop codon suppression (SCS)   151

t

TetCys motif   74ff.
thiol-activated toxin   63
transferrin (Tf)   18
– receptor 1 (TfR1)   127
transient confinement zone (TCZ)   55
trimethoprim (TMP) derivative   109ff.
– BODIPY   113
– fluorescent   113